Lecture Notes in Computer Science 15196

Founding Editors

Gerhard Goos
Juris Hartmanis

The series Lecture Notes in Computer Science (LNCS), including its subseries Lecture Notes in Artificial Intelligence (LNAI) and Lecture Notes in Bioinformatics (LNBI), has established itself as a medium for the publication of new developments in computer science and information technology research, teaching, and education.

LNCS enjoys close cooperation with the computer science R & D community, the series counts many renowned academics among its volume editors and paper authors, and collaborates with prestigious societies. Its mission is to serve this international community by providing an invaluable service, mainly focused on the publication of conference and workshop proceedings and postproceedings. LNCS commenced publication in 1973.

Klaus Drechsler · Cristina Oyarzun Laura ·
Moti Freiman · Yufei Chen · Stefan Wesarg ·
Marius Erdt

Editors

Clinical Image-Based Procedures

13th International Workshop, CLIP 2024
Held in Conjunction with MICCAI 2024
Marrakesh, Morocco, October 6, 2024
Proceedings

 Springer

Editors
Klaus Drechsler [ID]
Aachen University of Applied Sciences
Aachen, Germany

Moti Freiman [ID]
Technion – Israel Institute of Technology
Haifa, Israel

Stefan Wesarg [ID]
Fraunhofer Institute for Computer Graphics
Research IGD
Darmstadt, Germany

Cristina Oyarzun Laura
Fraunhofer Institute for Computer Graphics
Research IGD
Darmstadt, Germany

Yufei Chen
Tongji University
Shanghai, China

Marius Erdt [ID]
Singapore, Singapore

ISSN 0302-9743   ISSN 1611-3349 (electronic)
Lecture Notes in Computer Science
ISBN 978-3-031-73082-5   ISBN 978-3-031-73083-2 (eBook)
https://doi.org/10.1007/978-3-031-73083-2

Preface

The 13th International Workshop on Clinical Image-based Procedures: Towards Holistic Patient Models for Personalized Healthcare (CLIP 2024) was held in Marrakesh, Morocco, on October 6, 2024. As in previous years, it was organized in conjunction with the International Conference on Medical Image Computing and Computer Assisted Intervention (MICCAI 2024).

Continuing the long tradition of CLIP in translational research, the goal of our workshop is to bridge the gap between basic research methods and clinical practice. A key aspect of the applicability of these methods is the creation of holistic patient models, which is a crucial step towards personalized healthcare. As a matter of fact, the clinical picture of a patient is not exclusively composed of medical images. It is the combination of medical image data from multiple modalities with other patient data, such as omics, demographics, and electronic health records, that is most desirable. Since 2019, CLIP has placed special emphasis on this area of research.

Due to the high number of satellite events at MICCAI, CLIP 2024, like many other MICCAI workshops, was given space for a half-day event. We received 11 submissions, which is a good outcome considering the number of competing workshops. Based on the scores assigned by our reviewers, the quality of the received papers improved compared to previous years. All submitted papers were peer-reviewed by at least two experts, and nine papers were finally accepted for presentation at the workshop.

We would like to take this opportunity to thank MICCAI for providing the platform for our workshop. Furthermore, we also like to express our gratitude to our program committee members and authors who contributed to making CLIP 2024 a success.

August 2024

Klaus Drechsler
Cristina Oyarzun Laura
Moti Freiman
Yufei Chen
Stefan Wesarg
Marius Erdt

Organization

Organizing Committee

Drechsler, Klaus	Aachen University of Applied Sciences, Germany
Oyarzun Laura, Cristina	Fraunhofer Institute for Computer Graphics Research IGD, Germany
Freiman, Moti	Technion-Israel Institute of Technology, Israel
Chen, Yufei	Tongji University, China
Wesarg, Stefan	Fraunhofer Institute for Computer Graphics Research IGD, Germany
Erdt, Marius	Singapore, Singapore

Reviewers

Balaram, Shafa
Egger, Jan
Elsakka, Abdelrahman
Fragemann, Jana
Holmes, Tobias
Hoßbach, Martin
Karpate, Yogesh
Luijten, Gijs
Rheude, Tillmann
Yu, Yang
Zidowitz, Stephan

Contents

CloverNet – Leveraging Planning Annotations for Enhanced Procedural MR Segmentation: An Application to Adaptive Radiation Therapy

Francesca De Benetti[1(✉)], Yousef Yaganeh[1,2], Claus Belka[3,4,5],
Stefanie Corradini[3], Nassir Navab[1], Christopher Kurz[3], Guillaume Landry[3],
Shadi Albarqouni[6,7], and Thomas Wendler[1,8,9]

[1] Chair for Computer Aided Medical Procedures and Augmented Reality, Technische Universität München, Garching, Germany
[2] Munich Center for Machine Learning, Munich, Germany
[3] Department of Radiation Oncology, LMU University Hospital, LMU Munich, Munich, Germany
[4] German Cancer Consortium (DKTK), partner site Munich, a partnership between DKFZ and LMU University Hospital Munich, Munich, Germany
[5] Bavarian Cancer Research Center (BZKF), Munich, Germany
[6] Clinic for Interventional and Diagnostic Radiology, University Hospital Bonn, Bonn, Germany
[7] Helmholtz AI, Helmholtz Munich, Munich, Germany
[8] Department of Diagnostic and Interventional Radiology and Neuroradiology, University Hospital Augsburg, Augsburg, Germany
[9] Institute of Digital Health, University Hospital Augsburg, Neusaess, Augsburg, Germany

Abstract. In radiation therapy (RT), an accurate delineation of the regions of interest (ROI) and organs at risk (OAR) allows for a more targeted irradiation with reduced side effects. The current clinical workflow for combined MR-linear accelerator devices (MR-linacs) requires the acquisition of a planning MR volume (MR-P), in which the ROI and OAR are accurately segmented by the clinical team. These segmentation maps (S-P) are transferred to the MR acquired on the day of the RT fraction (MR-Fx) using registration, followed by time-consuming manual corrections. The goal of this paper is to enable accurate automatic segmentation of MR-Fx using S-P without clinical workflow disruption. We propose a novel UNet-based architecture, CloverNet, that takes as inputs MR-Fx and S-P in two separate encoder branches, whose latent spaces are concatenated in the bottleneck to generate an improved segmentation of MP-Fx. CloverNet improves the absolute Dice Score by 3.73% (relative +4.34%, p<0.001) when compared with conventional 3D UNet. Moreover, we believe this approach is potentially applicable to other longitudinal use cases in which a prior segmentation of the ROI is available.

Keywords: MRI · Radiation Therapy · MR-linac · Patient-specific Segmentation

© The Author(s), under exclusive license to Springer Nature Switzerland AG 2024
K. Drechsler et al. (Eds.): CLIP 2024, LNCS 15196, pp. 1–10, 2024.
https://doi.org/10.1007/978-3-031-73083-2_1

1   Introduction

Radiation Therapy (RT), an established treatment in oncology [14], utilizes primarily linear accelerator (linac) systems. Linacs with an integrated Magnetic Resonance (MR) scanner (MR-linac) have recently enabled MR-guided RT (MRgRT), a type of Online Adaptive RT (OART), which has superior imaging quality compared to conventional computed tomography (CT)-guided RT and allows real-time monitoring of the patient motion and delivered dose [13].

The initial step in the RT workflow in MRgRT is the acquisition of planning CT (CT-P) and MR (MR-P). The clinical team then manually annotates the MR-P with regions of interest (ROI) and organs at risk (OAR), resulting in a patient-specific planning segmentation map (S-P). CT-P and S-P are then used to generate the RT treatment plan (RT-TP).

At the beginning of each RT session, a so-called "fraction" MR (MR-Fx) is acquired to retrieve the current location of the ROI and OAR. This is needed because slight changes in the anatomy (i.e. due to how full the bowel, the stomach, and the bladder are) can happen and they can cause a mis-targeting of the RT. To avoid damaging the OAR or underdosing the ROI, the RT-TP needs to be updated. To do so, first, the MR-P and MR-Fx are registered, and then the obtained transformation is applied to S-P. However, current clinical registration tools are seldom sufficient to avoid misalignment of MR-P and MR-Fx. For this reason, the clinical team is still required to correct the registered segmentation map to generate the segmentation map of the current fraction (S-Fx), which perfectly matches MR-Fx. This is a time-consuming process, which can take up to ≈ 40% of the whole RT session [10]. Afterwards, the RT-TP is updated using S-Fx (so that it accounts for the anatomical changes), it is approved by the clinical team and the RT is delivered.

This work aims to develop an automatic deep-learning tool for segmenting ROI and OAR in MR-Fx for clinical practice. A key feature of the tool is using S-P as prior information to ensure reliable segmentation. Although conventional neural networks can segment ROI and OAR using a single input (i.e. only the medical image volume) [1,6], incorporating S-P introduces clinical expertise of the clinician segmenting MR-P, making the segmentation more robust and tailored to patient anatomy. Therefore, we designed a Convolutional Neural Network (CNN) that leverages S-P information to accurately segment MR-Fx.

The main contributions of this work are:

- We offer a pipeline for streamling the workflow of MRgRT which can be seamlessly integrated in the clinical workflow
- We propose a lightweight segmentation CNN, called CloverNet, able to incorporate prior information, without the need for a joint registration-segmentation or patient-specific training
- We benchmark our approach with a static 3D CNN and a patient-specific segmentation pipeline

1.1 Related Work

Propagating the planning contours to the MR-Fx it is being addressed by multiple research groups, in mainly two ways: patient-specific segmentation, or joint registration-segmentation.

Segmentation-based approaches rely on personalized annotations to train patient-specific segmentation CNNs. As an example, Kawula et al. [7] proposed to train a baseline UNet on multi-patient data and then fine-tune it using the MR-P of one specific patient, who was not included in the training set of the baseline. They tested their approach on a private MR-Fx dataset with 92 patients for the segmentation of bladder and rectum and reported a relative improvement ≈ +3% in Dice Score over the baseline when using the patient-specific fine-tuning. Li et al. [11] proposed a patient-specific CNN segmentation approach in which the DL model is trained first with the contours of the first MR-Fx, and then it is re-trained with the following ones. They reported an improvement in DSC over conventional registration methods of | 0.27 % in 6 in-house patients.

Some studies tried to simultaneously solve the problems of registration and segmentation in a multi-task learning fashion. For example, Zhou et al. [16] used a joint loss formulation for the training of one multi-modal registration CNN as well as two mono-modal segmentation CNNs, with the goal of using the predicted segmentations as an additional weak-supervision signal for the registration. In a mono-modality problem, Kohr et al. [8] proposed to use one registration and one segmentation CNN followed by cross-attention blocks. Finally, Elmahdy et al. [3] proposed to use two UNet-based networks, one for segmentation and one for registration, and applied a "cross-stitch network" approach [12] in which a dedicated learning-based unit is used to determine the amount of feature sharing between the two tasks.

The purely registration-based approaches for patient-specific segmentation [4,5,9] are outside of the scope of this work, but, in general, they focus on solving a global problem (i.e., the global alignment of the images) and not on the ROI. The use of additional information which specifies the ROI was shown to be helpful [5], but it would require the generation of the segmentation maps of both images beforehand.

As summary, joint registration-segmentation approaches often require complex architectures [3,8] and loss formulations [16], while the purely segmentation-based approaches can suffer from the limited amount of patient-specific data [11].

2 Methodology and Materials

This work serves as a proof of concept for incorporating prior patient-specific additional information into the learning process. We chose kidney segmentation as a showcase for our proposed method, but believe it can be applied to any anatomical structure with low deformation over time. For this study, both kidneys were segmented as a single class, though the methodology is adaptable to a multi-organ framework.

Baseline 1: 3D UNet. In this work, we use as first baseline architecture for conventional segmentation a 3D UNet [2] with 4 layers (with size [16, 32, 64, 128] and stride 2), single-channel input (MR-Fx), and double-channel output (S-Fx), in which the joint segmentation of both kidneys is returned in one-hot encoding format. We used PRELU as activation function inside the UNet and softmax after the last layer. In 3D UNet, S-P is not used in any way.

Baseline 2: TotalSegmentator-MRI. [1] Akinci D'Antonoli et al. [1] recently released the pre-trained weights for a nnUNet-based MR multi-organ segmentation task within the TotalSegmentator framework [15]. TotalSegmentator-MRI is trained on both MR (with different sequences) and CT volumes and can robustly segment major anatomical structures in MR images regardless of the sequence used [1]. Therefore, we utilized it as our second baseline for conventional segmentation.

Baseline 3: nnUNet. [6] We employed the nnUNet [6] approach to generate the third baseline of conventional segmentation. Among the possible training settings, we chose to train nnUNet for 250 epochs, to stay as close as possible to the training set-up of the proposed approach (see Sect. 2.3), as well as to use the suggested 5-fold cross-validation.

Baseline 4: Kawula et al. [7] As baseline for patient-specific segmentation, we followed the approach proposed by Kawula et al. [7]. First, we trained a baseline model (Kawula-B) on MR-P with cross-validation (see Sect. 2.2) using as architecture the baseline 3D UNet to ensure comparability. Afterwards, the patient-specific network (Kawula-PS) was obtained by finetuning the best-performing fold of Kawula-B using the MR-P of one test patient, and finally, it was tested on MR-Fx1 and MR-Fx5 of the same patient.

2.1 CloverNet Pipeline

Mimicking the current clinical workflow, the proposed pipeline begins with a rigid registration step that aligns MR-P and S-P with MR-Fx in a common coordinate space (MR-P$^{\mathcal{R}}$ and S-P$^{\mathcal{R}}$, respectively). It then employs a novel segmentation architecture, called CloverNet, which extracts relevant features from S-P$^{\mathcal{R}}$ to generate a more precise S-Fx (see Fig. 1). As previously noted, simply aligning S-P with MR-Fx is insufficient to address changes in patient anatomy. Therefore, CloverNet is designed to utilize S-P^{R} while correcting any misalignments.

Preliminary Rigid Registration. We rigidly registered MR-P to MR-Fx using the `SimpleITK` registration workflow, with the Normalized Cross Correlation as optimization metric. The transformation was then applied to S-P (S-P$^{\mathcal{R}}$) to bring it in the same coordinate space as MR-Fx. Following the consensus in the literature [3,4,9], MR-P is considered the "moving image" and MR-Fx the "fixed image".

Architecture. The proposed architecture extends the baseline 3D UNet by adding a new encoder branch $\mathcal{E}_S$, with exactly the same design as the 3D UNet encoder $\mathcal{E}_{MR}$. The latent space of $\mathcal{E}_S$ is concatenated to the latent space of $\mathcal{E}_{MR}$ and then fed to the bottleneck and afterwards to one common decoder $\mathcal{D}$ (see Fig. 1). The input of $\mathcal{E}_{MRI}$ and the output are the same as in the baseline UNet. The input of $\mathcal{E}_S$ is a two-channel tensor representing S-P$^{\mathcal{R}}$ in one-hot encoding.

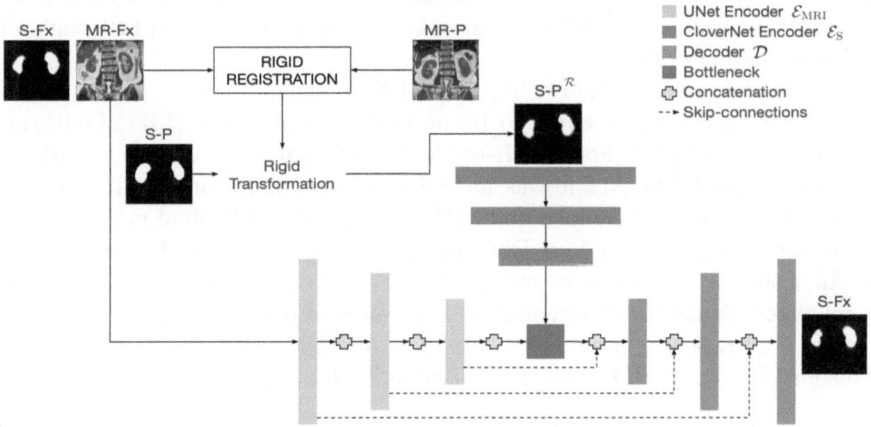

Fig. 1. Graphical illustration of the proposed pipeline with the CloverNet architecture.

2.2   Dataset

The data was acquired at the LMU University Hospital (Munich, Germany) between January 2020 and November 2022 from a total of 178 patients undergoing MRgRT for the treatment of various types of cancers in the abdomen. Informed written consent was obtained from all patients (LMU: ethics project number 20-291). The MR volumes were acquired with a 0.35 T MR-Linac (MRIdian, ViewRay Inc, Cleveland, Ohio) with a balanced steady-state free precession (bSSFP) sequence resulting in T2*/T1 contrast. Each patient received between 1 and 20 RT fractions. We considered only patients with MR-P, MR-Fx1 and/or MR-Fx5. Moreover, we curated the dataset by discarding patients with low-quality segmentation of the kidneys and those in which the field of view was not consistent among MR-P, MR-Fx1 and MR-Fx5. Afterwards, the dataset consists of a total of 65 patients and 104 volumes. The volumes were resampled to isotropic 1.5 mm spacing, but not resized. We applied 99th-percentile clipping to the MR volumes, and then rescaled them to $\in [0,1]$.

The dataset was split at a patient level: 11 patients, corresponding to 17 volumes, were used as test, 18 patients as validation and the remaining 36 patients as training set. We performed 3-fold cross-validation. Given that not all the patients have both MR-Fx1 and MR-Fx5, the size of the training set and the validation set was slightly different depending on the fold ($\in [57, 59]$ and $\in [28, 30]$, respectively).

2.3   Experimental Set-Up: 3D UNet and CloverNet

The networks were trained in a patch-based approach (with patch size equal to [112, 112, 112]) to limit memory usage when processing large 3D volumes. The selection of the patches was performed by `RandCropByPosNegLabeld` from the `monai` library, and 75% of the samples were chosen to have a foreground voxel in the middle of the patch. Following Kawula et al. [7], we applied as data augmentation `RandAffine`, `RandZoom`, `RandBiasField` from the `monai` library, and `RandomNoise` from the `torchio` library. All augmentation was applied randomly with a probability of 0.5.

The models were trained using ADAMW as optimizer for a maximum 150 epochs, with a batch size equal to 10, an initial learning rate (LR) of 0.01, and instance normalization. For 3D UNet the LR was halved at epochs 30, 40, and 50, whereas for CloverNet it was halved at epochs 5, 10, and 20 to take into account the faster convergence due to the use of S-P$^{\mathcal{R}}$. The final model was the one of the epoch with the best DSC on the kidneys on the validation set.

As a post-processing step, we applied to all the predictions `FillHoles` and `KeepLargestConnectedComponent` as implemented in the `monai` library.

The Dice Loss was used as loss function, and Dice Score (DSC [%]) and the Hausdorff Distance (HD [mm]) were used as evaluation metrics.

Table 1. Evaluation of the presented approaches (mean (median) $\pm$ standard deviation) on the test set. For the approaches with N-fold cross-validation, the average metric of the N folds in the test set is reported. The statistical significance against 3D UNet (*: p-value < 0.05, **: p-value < 0.001) is reported for the patient-specific 3D UNet-based approaches.

Architecture	DSC [%]	HD [mm]
Initial	37.16 (27.89) $\pm$ 21.38	23.84 (23.89) $\pm$ 8.40
After Rigid Registration	72.12 (73.45) $\pm$ 6.09	14.41 (13.49) $\pm$ 4.77
3D UNet	86.00 (86.92) $\pm$ 3.79	12.39 (11.47) $\pm$ 4.23
TotalSegmentator-MRI [1]	83.25 (83.92) $\pm$ 0.02	11.61 (11.04) $\pm$ 1.48
nnUNet [6]	**92.93 (93.24) $\pm$ 1.24**	**8.00 (8.24) $\pm$ 2.47**
Kawula-B [7]	84.87 (85.36) $\pm$ 3.61	13.11 (12.09) $\pm$ 3.22
Kawula-PS [7]	88.33 (88.75) $\pm$ 1.77 *	11.10 (10.72) $\pm$ 3.87
CloverNet	**89.73 (90.23) $\pm$ 1.81** **	**10.58 (10.28) $\pm$ 4.07**

3   Results

First, we performed a naive approach by simply propagating the S-P to MR-Fx. Without applying any registration, this resulted in an average DSC in the test dataset of 37.16$\pm$21.38% ($\in$ [9.16, 75.52]%), whereas after the rigid registration the average DSC was 72.12$\pm$6.09 % ($\in$ [54.92, 78.39]%).

Among the conventional segmentation approaches, namely 3D UNet, TotalSegmentator-MRI [1], nnUNet [6] and Kawula-B [7], the best performing one is nnUNet, with a DSC of 92.93 ± 1.24 % and an HD of 8.00 ± 2.47 mm (see Table 1).

On the other hand, among the patient-specific approaches, CloverNet is superior to Kawula-PS [7], with +1.4 % DSC and -0.52 mm HD, being statistically significantly different in DSC with p-value < 0.05.

The well-known superiority of nnUNet [6] was demonstrated also in our task, where it achieved an improvement ∈ [+6.93, +9.68] % DSC and ∈ [-5.11, -3.61] mm HD against the other conventional segmentation approaches.

However, the performance of the proposed CloverNet should not be directly compared against to nnUNet because CloverNet does not employ the automatic fingerprinting pipeline of nnUNet. On the other hand, 3D UNet and Clover-Net can be fairly compared because they only differ in the architecture design. Indeed, the performance of CloverNet was superior to the one of the baseline UNet by +3.73 % DSC and -1.81 mm HD, and the difference in DSC was statistically significant with p-value < 0.001 (see Fig. 2).

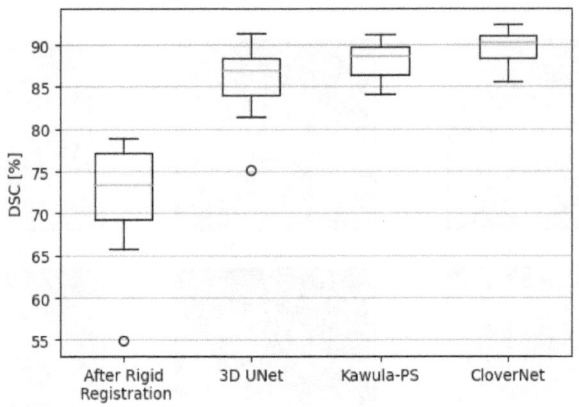

Fig. 2. Box plot representation of the DSC [%] of the UNet-based approaches.

4 Discussion

The superiority of nnUNet [6] with respect to both UNet-based patient-specific approaches, Kawula-PS [7] and the proposed CloverNet, is not surprising, given its well-known and widely reported performance. This motivates us to continue developing a nnUnet-based patient-specific automatic segmentation tool. Nonetheless, we continue the discussion focusing only on the UNet-based approaches to ensure fair comparability.

Our proposed framework, the CloverNet pipeline, represents a clinically aligned solution for incorporating patient-specific prior information into RT.

By leveraging readily available data during MRgRT, we avoid disruptions to the clinical workflow. The statistical rationale behind the CloverNet pipeline lies in its simplicity, seamless integration, and potential to enhance personalized outcomes. As clinicians continue their established practices, they can confidently utilize our framework to speed up RT-TP and improve patient care without any additional burden or complexity.

Given that CloverNet partially relies on S-P to segment MR-Fx, high-quality annotation in S-P is crucial. Clinical practice shows that annotating physicians have more time during the planning stage, ensuring this quality. Training exclusively on S-P (as in Kawula-B) would reduce the amount of available data, while training solely on S-Fx (as in 3D UNet) would reduce the annotation quality due to time constraints during RT delivery. Consequently, patient-specific segmentation methods like Kawula-PS and CloverNet outperform conventional non-patient-specific approaches.

Although CloverNet was only slightly superior to Kawula-PS [7] (see Fig. 3), it has higher clinical applicability because it requires only one training phase, and the same model can be applied to multiple patients without the need of time- and memory-consuming patient-specific fine-tuning.

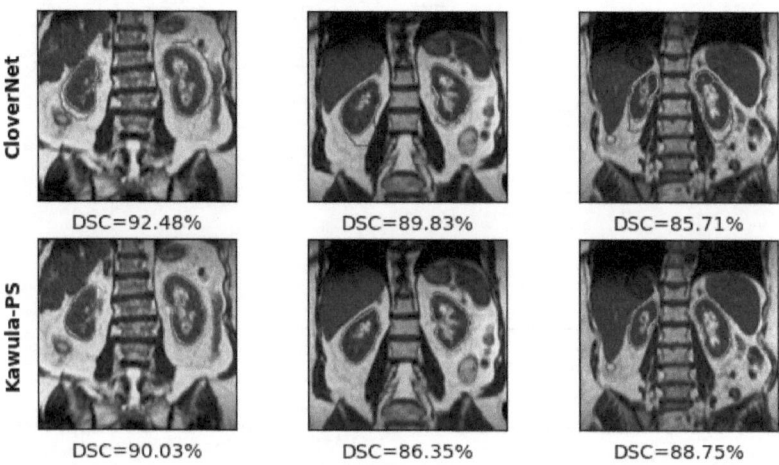

Fig. 3. Examples of segmentation results for the two patient-specific approaches and three selected patients (left to right), including the DSC [%], the contour of the ground truth (red) and prediction (green), and of S-P^R (black), when relevant (Color figure online).

5   Conclusion

In this work, we address the problem of segmentation of MR-Fx in the case in which a previous annotation of the ROI is available, as required for MRgRT. The proposed architecture, CloverNet, includes a dedicated encoder to encode the spatial information of the S-P$^\mathcal{R}$ and then uses it in the segmentation of MR-Fx. We showed that the proposed approach was better than a conventional 3D UNet baseline, and therefore we proved that the information coming from S-P$^\mathcal{R}$ was beneficial to segment MR-Fx.

The design of the proposed CloverNet is superior to other comparable approaches for MR-Fx segmentation for multiple reasons. First of all, CloverNet only performs rigid registration as a pre-processing step and relies on the segmentation network to fine-tune the initial segmentation. This ensures the realism of the S-P$^\mathcal{R}$ and no folding artifacts, which are often reported in the purely registration-based approaches [5,9]. Second, its design is simple and it does not require complex architectures or loss formulations [3,8,16]. Finally, compared to other segmentation-based approaches for patient-specific segmentation [7,11], CloverNet does not require additional training for each new patient or fraction, giving it a significant advantage in terms of time and the need for computational resources. All these aspects make CloverNet more suitable for clinical use.

While in this work we concentrate on MRgRT, the proposed pipeline could be used in other longitudinal segmentation cases, for example where registration-based approaches struggle. Indeed, CloverNet only requires a clinical scenario where a planning segmentation is available at the time a procedural medical image is acquired and needs to be segmented. Moreover, we hypothesize that the planning segmentation could be based on another modality, given that the prior information is only a binary mask.

Personalized medicine is one of the emerging fields in clinical practice, and patient-specific approaches are essential for its implementation. In segmentation, non-personalized approaches with great performance [6] are already available. However, they are not patient-specific and can not make use of previous knowledge on the patient to perform the segmentation, which is sub-optimal. Using patient-specific approaches, particularly in scenarios like MRgRT where segmentation accuracy is crucial, would reduce the time needed to correct the automatically-generated segmentations. In this perspective, CloverNet integrates easily the prior segmentation, and guarantees robust results, enabling seamless integration in clinical routine.

Acknowledgements. This work was partially funded by the German Research Foundation (DFG, grant 469106425 - NA 620/51-1).

References

1. D'Antonoli, T.A., et al.: TotalSegmentator MRI: sequence-independent segmentation of 59 anatomical structures in MR images. arXiv preprints arXiv:2405.19492 (2024)
2. Çiçek, Ö., Abdulkadir, A., Lienkamp, S.S., Brox, T., Ronneberger, O.: 3D U-Net: learning dense volumetric segmentation from sparse annotation. In: Medical Image Computing and Computer-Assisted Intervention–MICCAI 2016: 19th International Conference, Athens, Greece, 2016, Proceedings, Part II 19 (2016)

3. Elmahdy, M.S., et al.: Joint registration and segmentation via multi-task learning for adaptive radiotherapy of prostate cancer. IEEE Access **9**, 95551–95568 (2021)
4. Eppenhof, K.A., et al.: Fast contour propagation for MR-guided prostate radiotherapy using convolutional neural networks. Med. Phys. **47**(3), 1238–48 (2020)
5. Hemon, C., et al.: Contour-guided deep learning based deformable image registration for dose monitoring during CBCT-guided radiotherapy of prostate cancer. J. Appl. Clin. Med. Phys. **24**(8), e13991 (2023)
6. Isensee, F., Jaeger, P.F., Kohl, S.A., Petersen, J., Maier-Hein, K.H.: nnU-Net: a self-configuring method for deep learning-based biomedical image segmentation. Nat. Methods **18**(2), 203–11 (2021)
7. Kawula, M., et al.: Prior knowledge based deep learning auto-segmentation in magnetic resonance imaging-guided radiotherapy of prostate cancer. Phys. Imag. Radiat. Oncol. **28**, 100498(2023)
8. Khor, H.G., Ning, G., Sun, Y., Lu, X., Zhang, X., Liao, H.: Anatomically constrained and attention-guided deep feature fusion for joint segmentation and deformable medical image registration. Med. Image Anal. **88**, 102811 (2023)
9. Kolenbrander, I.D., et al.: Deep-learning-based joint rigid and deformable contour propagation for magnetic resonance imaging-guided prostate radiotherapy. Med. Phys. **51**(4), 2367–77 (2024)
10. Landry, G., Kurz, C., Traverso, A.: The role of artificial intelligence in radiotherapy clinical practice. BJR Open **5**(1), 20230030 (2023)
11. Li, Z., et al.: Patient-specific daily updated deep learning auto-segmentation for MRI-guided adaptive radiotherapy. Radiother. Oncol. **177**, 222–230 (2022)
12. Misra, I., Shrivastava, A., Gupta, A., Hebert, M.: Cross-stitch networks for multi-task learning. In: Proceedings of the IEEE conference on computer vision and pattern recognition (2016)
13. Ng, J., et al.: MRI-LINAC: a transformative technology in radiation oncology. Front. Oncol. **13**, 1117874 (2023)
14. Shepherd, M., et al.: A scoping review of advanced practice in online adaptive radiotherapy: educational needs and training for evidence and opportunity building. J. Med. Imag. Radiat. Sci. **54**(4), S6 (2023)
15. Wasserthal, J., et al.: TotalSegmentator: robust segmentation of 104 anatomic structures in CT images. Radiol. Artif. Intell. **5**(5) (2023)
16. Zhou, Z., et al.: macJNet: weakly-supervised multimodal image deformable registration using joint learning framework and multi-sampling cascaded MIND. Biomed. Eng. Online **22**(1),(2023). https://doi.org/10.1186/s12938-023-01143-6

Enhancing Image Classification in Small and Unbalanced Datasets Through Synthetic Data Augmentation

Neil de la Fuente[1,2(✉)], Mireia Majó[1,2], Irina Luzko[3], Henry Córdova[3], Gloria Fernández-Esparrach[3], and Jorge Bernal[1,2]

[1] Computer Vision Center, Barcelona, Spain
{neil.delafuente,mireia.majo,jorge.bernal}@autonoma.cat
[2] Universitat Autònoma de Barcelona, Barcelona, Spain
[3] Hospital Clinic, Barcelona, Spain
{luzko,hcordova,mgfernan}@clinic.cat

Abstract. Accurate and robust medical image classification is a challenging task, especially in application domains where available annotated datasets are small and present high imbalance between target classes. Considering that data acquisition is not always feasible, especially for underrepresented classes, our approach introduces a novel synthetic augmentation strategy using class-specific Variational Autoencoders (VAEs) and latent space interpolation to improve discrimination capabilities. By generating realistic, varied synthetic data that fills feature space gaps, we address issues of data scarcity and class imbalance. The method presented in this paper relies on the interpolation of latent representations within each class, thus enriching the training set and improving the model's generalizability and diagnostic accuracy.

The proposed strategy was tested in a small dataset of 321 images created to train and validate an automatic method for assessing the quality of cleanliness of esophagogastroduodenoscopy images.

By combining real and synthetic data, an increase of over 18% in the accuracy of the most challenging underrepresented class was observed. The proposed strategy not only benefited the underrepresented class but also led to a general improvement in other metrics, including a 6% increase in global accuracy and precision.

Keywords: Synthetic Data Augmentation · Variational Autoencoder · Esophagogastroduodenoscopy Image Classification · Image Classification

1 Introduction

Gastric cancer (GC) is the 5th most common cancer worldwide and there were more than 1 million new cases of GC reported in 2020. Esophagoduodenoscopy (EGD) is the gold standard method for the diagnosis of GC: several studies show that the detection of GC at earlier stages has a clear impact in the decrease of the

K. Drechsler et al. (Eds.): CLIP 2024, LNCS 15196, pp. 11–21, 2024.
https://doi.org/10.1007/978-3-031-73083-2_2

mortality (hazard ratio [HR] 0.51) [1,2]. Nevertheless, up to 10% of the cancers are missed during the exploration, with a clear impact on patient's survival rate [3]. Poor mucosal visualization is one of the factors that can negatively affect the diagnostic accuracy of gastric cancer.

For this reason, the degree of cleanliness and the quality of gastric mucosa visibility are of paramount importance. However, no broadly accepted cleanliness scale for the upper gastrointestinal tract (UGI) has been uniformly accepted and used in routine practice. Two scales have been recently published: POLPREP [4] and Barcelona scale [5]. Both evaluate the level of cleanliness in the esophagus, stomach and duodenum. They differ on the number of levels (4 for POLPREP, 3 for Barcelona scale) and in the degree of evaluation detail: Barcelona scale further divides stomach by segments (fundus, corpus and antrum).

However, these scales are prone to a certain degree of subjectivity. To cope with this, and following other methods already developed to assist clinicians in similar tasks [6], there is room for AI systems that can provide an objective assessment of the degree of UGI cleanliness by an automatic classification of EGD images. The benefits of such a system are clear: if clinicians can be sure of those cases when gastroscopies are inappropriate due to insufficient cleanliness, they can make a recommendation to repeat the exploration. In the opposite case, where the UGI is clean, unnecessary repetitions can be avoided with the consequent saving of scarce economic resources.

The main technical challenge in medical image classification comes from the limited size and imbalance of available datasets. This limitation reflects the real-world shortage of annotated medical images and uneven distribution of pathological findings, making it difficult to develop robust models with traditional deep learning which usually requires large, balanced datasets. Additionally, the detailed nature of EGD images, which needs accurate identification of different levels of cleanliness, adds to the challenge. The scarcity of significant features in smaller datasets can result in model biases or underperformance. Overcoming these challenges requires innovative approaches that improve data diversity and representation, enhancing model's ability to generalize in real-world scenarios.

The key contributions of our work include:

- **Use of class-Specific variational autoencoders (VAEs):** By generating synthetic images through latent representation interpolation within classes, we can expand the feature space and directly address class imbalance.
- **Focused enrichment of feature space:** Our technique fills gaps in the feature space with realistic synthetic images, improving training effectiveness and model sensitivity to critical subtle features for accurate classification.
- **Proof of the versatility of our approach across architectures:** We demonstrate the benefits of our methodology across two prominent image classification architectures such as EfficientNet-V2 [20] and ResNet-50.

2   Related Work

The previously mentioned scarcity of annotated medical datasets has led to novel strategies for data augmentation in image classification.

Garay-Maestre et al. [7] exploited Variational Autoencoders (VAEs) to generate synthetic samples, demonstrating how synthetic data, when combined with traditional augmentation methods, can improve the robustness and performance of machine learning models. Following this line of thought, Auzine et al. [8] applied Generative Adversarial Networks (GANs) with conventional augmentation to enhance the accuracy of deep learning architectures, such as ResNet50 [17] and VGG16 [18], on endoscopic esophagus imagery. Further advancing the field, Zhou et al. [19] introduced 'Diffusion Inversion', a technique for creating synthetic data by manipulating the latent space of pre-trained diffusion models to achieve comprehensive data manifold coverage and improved generalization.

Liu et al. [9] investigated data augmentation via latent space interpolation for image classification, showing significant improvements in model performance. Oring [10] and Cristovao et al. [11] also explored the use of VAEs for generating in-between images through latent space interpolation, emphasizing the potential of this approach for enhancing data diversity. Moreno-Barea et al. [12] and Elbattah et al. [13] focused on improving classification accuracy using data augmentation techniques on small datasets, highlighting the effectiveness of synthetic data in addressing class imbalance. Wan et al. [14] specifically addressed imbalanced learning through VAE-based synthetic data generation, demonstrating its potential to improve model performance on imbalanced datasets.

While data augmentation has been widely used in several research domains there is no work, to the best of our knowledge, that applies this technique to assess the degree of cleanliness of EGD images. Nevertheless there are works applied to similar images, such as the work of Nam et al. [16], which use InceptionResnetV2 to classify wireless capsule endoscopy images to determine the degree of mucosa visualization or the work of Zhu et al. [15], which applies a compact convolutional neural network with 2 Densenet layers to label bowel preparation in colonoscopy images according to Boston Bowel Preparation Scale.

While some of the previously mentioned studies have demonstrated the potential of generative models for data augmentation, our approach specifically targets class imbalance in EGD image classification by leveraging class-specific VAEs. By generating realistic synthetic images for each class, we aim to fill gaps in the feature space, thereby enhancing the training process and improving sensitivity to subtle features. This synthetic data augmentation methodology stands out by directly addressing the challenges posed by small and unbalanced datasets, particularly for EGD images, and notably improving performance.

3   Methodology

Variational Autoencoders (VAEs) [21] represented a groundbreaking shift in the generation of synthetic data by providing a probabilistic approach to data encoding and decoding. A VAE is composed of an encoder, for translating input data into a latent space representation, and a decoder, for reconstructing data from this latent space. The encoder function, denoted as $q_\phi(z|x)$, maps an input x to a latent space representation z, parameterized by ϕ.

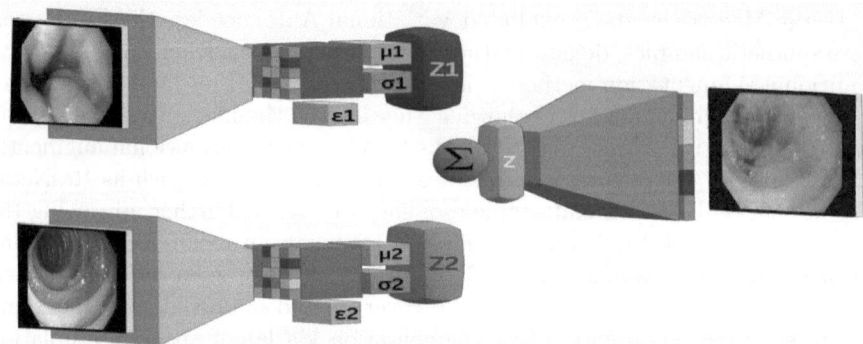

Fig. 1. The encoding and decoding process in VAEs for synthetic image generation via latent vector interpolation.

This process introduces a stochastic element by generating a distribution characterized by mean μ and variance σ, rather than a fixed point for the latent variables. This distribution allows for the sampling of new data points from the latent space, using the reparameterization trick: $z = \mu + \sigma \odot \epsilon$, where ϵ is an element-wise product with a random noise vector.

The decoder, denoted as $p_\theta(x|z)$ and parameterized by θ, reconstructs the data from the latent space representation. The objective of training a VAE is to minimize the loss function $\mathcal{L}(\theta, \phi; x)$, which is a combination of the negative log-likelihood of the reconstructed data and the Kullback-Leibler (KL) divergence, promoting an effective balance between data reconstruction fidelity and distribution approximation:

$$\mathcal{L}(\theta, \phi; x) = -\mathbb{E}_{q_\phi(z|x)}[\log p_\theta(x|z)] + \mathrm{KL}(q_\phi(z|x)\|p(z)) \tag{1}$$

The flexibility in generating new data points through this probabilistic framework makes VAEs particularly suitable for tasks like medical image augmentation, where capturing the diversity of pathological features is crucial.

The capabilities of VAEs are leveraged to generate synthetic images tailored for each class within the dataset. By training class-specific autoencoders, the unique characteristics and nuances of each class are captured in the latent space.

Post-training, the VAE synthesizes new images by performing an interpolation between the latent representations of two images within the same class, with the process illustrated in Fig. 1. This interpolation is achieved by computing a weighted sum of their latent vectors z_1 and z_2, yielding a new latent representation, denoted as z_{interp}:

$$z_{interp} = \alpha z_1 + (1 - \alpha)z_2 \tag{2}$$

where α corresponds to the interpolation weight $[0, 1]$. This parameter modulates the contribution of each original image to the synthesized image. The resulting latent vector z_{interp} is then decoded to generate a new synthetic image. This new image merges characteristics of the parent images while maintaining the class's defining features, effectively enriching the dataset's diversity and quantity.

By integrating additional synthetic images for each class, this iterative and monitored approach not only augments the dataset but also introduces meaningful and realistic variations which enrich the feature space.

The incorporation of these synthetic images into the training process is designed to enhance the classifiers' performance, particularly by improving its understanding of the underlying distribution of the underrepresented classes, thereby ensuring a more comprehensive representation of each class's feature space and subsequently elevating model accuracy.

4   Experimental Setup

4.1   Dataset

Our study utilized a dataset comprising 321 esophagogastroduodenoscopy (EGD) images, capturing various stomach regions including the esophagus, duodenum, antrum, body, and fundus. Images were labelled into different categories according to different degree of stomach cleanliness by seven clinicians following the definitions of the Barcelona scale.

The consensus among the experts determined the final class assignment: 1) class 0, which corresponds to images with presence of non aspirable solid or semisolid particles, bile or foam preventing from clear mucosa visualization; 2) class 1, which corresponds to images with small amount of semisolid particles, bile or foam and 3) class 2, which comprises images without any kind of rest, allowing a complete visualization of the mucosa. We show in Fig. 2 an example of some of the images in the dataset.

With respect to class distribution within the dataset, class 0 was represented with 65 images, class 1 with 165 and class 2 with 91. The dataset was split in a standard 80–20 fashion for training/validation and testing.

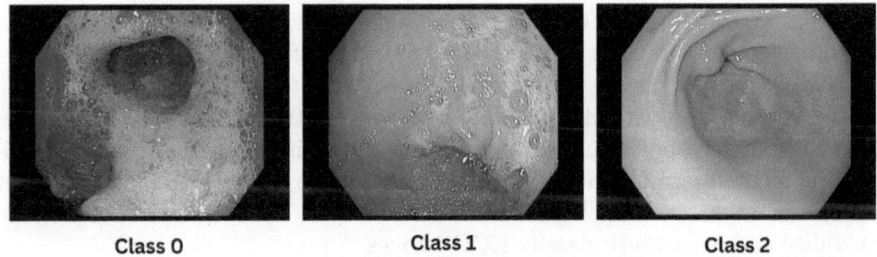

Class 0   Class 1   Class 2

Fig. 2. Sample image for each of the classes, labelled according to Barcelona scale.

4.2 Metrics

To assess model performance, a suite of metrics was employed, including overall accuracy, precision, recall, and F1-score, complemented by class-specific measures for a thorough analysis. The evaluation concentrated on the efficacy of several augmentation strategies, encompassing traditional methods such as rotations and mirroring, as well the proposed approach involving synthetic data generation via VAEs. Results were tabulated to compare the impact of these techniques on model performance.

4.3 Implementation Details

Synthetic Data Generation: The synthetic data was generated using class-specific Variational Autoencoders (VAEs). Each VAE was trained separately for each class, capturing the unique characteristics of the respective class. The architecture of the VAE consisted of an encoder and decoder, with the encoder mapping input images to a latent space and the decoder reconstructing images from the latent space. We selected a latent space dimension of 256 based on empirical analysis, and the VAEs were trained for 1000 epochs with a learning rate of 0.0001, using the Adam optimizer. To generate synthetic images, we performed latent space interpolation between pairs of latent vectors within the same class, ensuring realistic and varied synthetic samples. We experimented with different quantities of synthetic images, ultimately finding that adding 300 extra images per class yielded the best performance, hence, results shown in Sect. 5 include 300 synthetic images per class.

Classification: For the classification task, we employed two well-known architectures: EfficientNet-V2 and ResNet-50. Both pretrained models were implemented and fine-tuned using PyTorch. The training process involved a batch size of 24 images and an initial learning rate set to 5×10^{-4}. The learning rate was dynamically adjusted based on performance plateaus, with a reduction factor of 10 upon stagnation in validation loss improvement. To ensure reproducibility and consistency across experiments, a random seed was set to 42.

Multiple configurations were explored for both EfficientNet-V2 and ResNet-50 architectures, including training on real data with and without traditional augmentations, and extending the dataset with synthetically generated images through VAE interpolation. Synthetic images were integrated into both classically augmented and non-augmented training sets to assess the combined effect of classical and synthetic augmentation techniques on the models' ability to generalize and accurately classify EGD images.

5 Results

Experimental results, presented in Table 1, remark the efficacy of incorporating synthetic data augmentation via class-specific VAEs in enhancing classification

Table 1. Classification performance of EfficientNet-V2 and ResNet-50 models with various data augmentation strategies.

Model	Data	Overall Acc.	Overall Prec.	Overall Rec.	Overall F1	Class-0 Acc.	Class-1 Acc.	Class-2 Acc.
EfficientNet-V2	Real No Aug.	85.94	86.18	86.4	85.3	92.81	64.05	96.85
	Real with Aug.	87.5	87.51	87.5	87.46	81.43	74.56	97.10
	Real + Gen No Aug.	89.06	88.96	88.88	88.74	**92.85**	72.22	97.15
	Real + Gen with Aug.	**92.19**	**92.27**	**91.03**	**91.93**	92.23	**82.06**	**98.73**
ResNet-50	Real No Aug.	82.81	82.84	82.81	81.41	**92.38**	52.18	**96.72**
	Real with Aug.	84.38	84.3	84.38	84.38	88.89	68.89	94.15
	Real + Gen No Aug.	86.02	85.8	85.94	85.16	92.00	66.91	96.55
	Real + Gen with Aug.	**89.49**	**88.93**	**89.06**	**88.74**	92.15	**75.12**	96.61

performance. A comparative analysis across different models and augmentation strategies reveals that the addition of VAE-generated synthetic images leads to substantial improvements in overall accuracy, precision, recall, and F1-scores.

Notably, the most pronounced gains are observed in the accuracy metrics for the most challenging underrepresented class; class 1, where the conventional data pool is limited. The proposed strategy achieves the best performance in the majority of the experiments, being a very close second in the remaining cases.

The impact of different augmentation techniques on Class-1 accuracy is further highlighted in Fig. 3. The results confirm that synthetic data augmentation, particularly when combined with traditional augmentation techniques, substantially improves the robustness of classification models.

EfficientNet-V2, when trained with both real and synthetically augmented data, shows an overall accuracy boost from 85.94% to 92.19%, and a significant increase in Class-1 accuracy from 64.05% to 82.06%. Similarly, ResNet-50's performance escalates from 82.81% to 89.49% in overall accuracy, with Class-1 accuracy rising from 52.18% to 75.12%. These enhancements suggest that

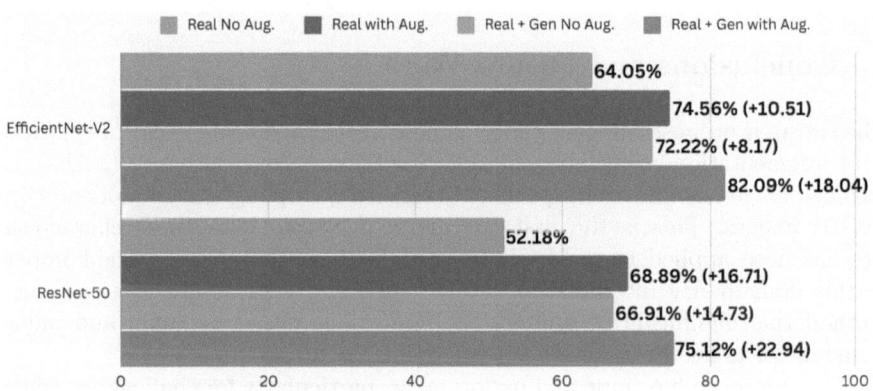

Fig. 3. Comparison of Class 1 Accuracy Across Different Augmentation Techniques and Classifiers. Improvement points are with respect to the Real No Augmentation bar for each model.

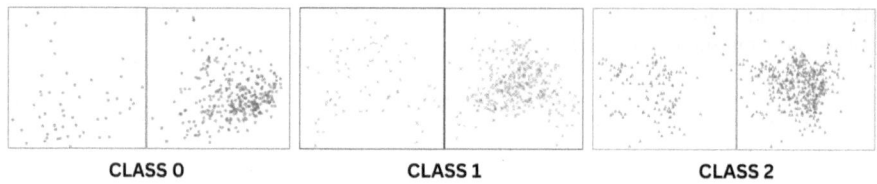

Fig. 4. Expansion of feature space for each EGD image class post-augmentation. X and Y axes represent PCA features 1 and 2 respectively.

synthetic data not only supplements the training set but also instills a better understanding of the feature space associated with each class.

Importantly, the augmented data appears to guide the model towards a more detailed comprehension of the subtle distinctions within the EGD image classes. This is critical for clinical applications where the differentiation between varying levels of cleanliness directly impacts the diagnostic process and subsequent patient care. Therefore the use of VAEs for data augmentation could suppose an advancement for medical imaging fields struggling with data constraints.

Figure 4 represents the data distribution for each class before and after synthetic augmentation. The original sparse distribution of each of the classes, as seen on the left side of each class's panel, becomes notably denser on the right side, following the application of VAE-based latent vector interpolation. This visual enhancement of the feature space is especially significant for Class 1, the primary focus of our study, where the augmented data points fill previously underpopulated regions, indicating a more balanced representation post-augmentation.

These visual findings align with the quantitative improvements in classification performance, confirming the value of VAE-based synthetic data in addressing class imbalance and enhancing model training for medical diagnostics.

6   Conclusions and Future Work

The research proposed in this paper demonstrates the effectiveness of synthetic data augmentation using class-specific Variational AutoEncoders (VAEs) for medical image classification, specifically targeting esophagogastroduodenoscopy (EGD) images. This is the first time such a system with these characteristics has been applied to the EGD imaging field, marking a significant impact in this domain. By interpolating latent representations within classes, a new method that significantly counters the limitations posed by small and unbalanced datasets has been developed and validated.

This approach has improved performance, particularly for challenging underrepresented classes, by effectively filling feature space gaps and achieving a more uniform dataset distribution. The success of this approach is demonstrated across two distinct architectures, EfficientNet-V2 and ResNet-50, showing its adaptability and the broad applicability of synthetic data augmentation in improving model classification capabilities.

Furthermore, the study explored the compounded benefits of combining traditional augmentation techniques with synthetic data augmentation, revealing a notable enhancement in the models' ability to generalize and accurately classify EGD images, discovering a synergistic effect.

The proposed work represents a significant step forward in utilizing AI for medical diagnostics, particularly by employing a methodologically innovative approach to synthetic data generation. By focusing on class-specific latent representation interpolations, it provides a scalable solution to the persisting problem of data scarcity and imbalance in medical imaging. This innovative methodology has set a precedent in the EGD imaging field, paving the way for its potential application in other medical imaging domains.

Despite these promising results, several limitations should be noted. The synthetic images generated by VAEs, while effective, may not capture all the nuances of real medical images, potentially leading to some biases in the training process. Additionally, the study was conducted on a relatively small dataset, which may limit the generalizability of the findings.

This contribution lays the groundwork for future explorations into more sophisticated synthetic data generation methods and their application across various domains within medical image analysis. A compelling direction for this research could involve adopting latent diffusion models (LDMs), well known for their capacity to generate high-quality, realistic images, to augment the diversity and authenticity of synthetic medical images. This approach, coupled with assessing the impact of such augmentation techniques on larger and more diverse medical image datasets, could significantly advance the scalability, robustness and applicability of these methods.

Finally, refining the interpolation techniques for synthetic image creation to achieve more precise and clinically relevant datasets remains a critical area for development. Future efforts could also focus on understanding how synthetic data influences model interpretability and reliability in real-world clinical environments, aiming to not only elevate classification accuracy but also improve the trust and efficacy of diagnostic models in medical practice.

Acknowledgements. This work was supported by the following Grant Numbers: PID2020-120311RB-I00 and RED2022-134964-T and funded by MCIN-AEI/10.13039/501100011033.

Disclosure of Interests. The authors have no competing interests to declare.

References

1. Ezoe, Y., Muto, M., et al.: Magnifying narrowband imaging is more accurate than conventional white-light imaging in diagnosis of gastric mucosal cancer. Gastroenterology **141**(6), 2017–2025 (2011)
2. Guillena, P.G.D., Alvarado, V.J.M., et al.: Gastric cancer missed at esophagogastroduodenoscopy in a well-defined Spanish population. Dig. Liver Dis. **51**(8), 1123–1129 (2019)
3. Tsukuma, H., Oshima, A., et al.: Natural history of early gastric cancer: a non-concurrent, long term, follow up study. Gut **47**(5), 618–621
4. Romańczyk, M., Ostrowski, B., et al.: Scoring system assessing mucosal visibility of upper gastrointestinal tract: the POLPREP scale. J. Gastroenterol. Hepatol. **37**(1), 164–168 (2022)
5. Córdova, H., Barreiro-Alonso, E., et al.: Applicability of the Barcelona scale to assess the quality of cleanliness of mucosa at esophagogastroduodenoscopy. Gastroenterol. Hepatol. **47**, 246–252 (2024)
6. Haithami, M.S., et al.: Automatic bowel preparation assessment using deep learning. In: International Conference on Pattern Recognition, pp. 574–588 (2022)
7. Garay-Maestre, U., Gallego, A.-J., Calvo-Zaragoza, J.: Data augmentation via variational auto-encoders. In: Vera-Rodriguez, R., Fierrez, J., Morales, A. (eds.) CIARP 2018. LNCS, vol. 11401, pp. 29–37. Springer, Cham (2019). https://doi.org/10.1007/978-3-030-13469-3_4
8. Auzine, M., et al.: Endoscopic image analysis using deep convolutional GAN and traditional data augmentation. In: Proceedings of the International Conference on Electronics, Communications and Control Engineering (ICECCME), pp. 1–6. IEEE (2022)
9. Liu, X., et al.: Data augmentation via latent space interpolation for image classification. In: Proceedings of the International Conference on Pattern Recognition (ICPR) (2018)
10. Oring, A.: Autoencoder image interpolation by shaping the latent space. arXiv preprint arXiv:2008.01487 (2020)
11. Cristovao, P., et al.: Generating in-between images through learned latent space representation using variational autoencoders. IEEE Access **8**, 149456–149467 (2020)
12. Moreno-Barea, F.J., et al.: Improving classification accuracy using data augmentation on small data sets. Expert Syst. Appl. **161**(15), 113696 (2020)
13. Elbattah, M., et al.: Variational autoencoder for image-based augmentation of eye-tracking data. J. Imag. **7**(5), 83 (2021)
14. Wan, Z., Zhang, Y., He, H.: Variational autoencoder based synthetic data generation for imbalanced learning. In: IEEE Symposium on Computational Intelligence (SSCI), Honolulu, HI, USA, 2017, pp. 1–7 (2017). https://doi.org/10.1109/SSCI.2017.8285168
15. Zhu, Y., Yiwen, X., et al.: A CNN-based cleanliness evaluation for bowel preparation in colonoscopy. In: 12th International Congress on Image and Signal Processing, BioMedical Engineering and Informatics (CISP-BMEI), pp. 1–5. IEEE (2019)
16. Nam, J.H., et al.: Development of a deep learning-based software for calculating cleansing score in small bowel capsule endoscopy. Sci. Rep. **11**(1) (2021)
17. He, K., et al.: Deep residual learning for image recognition. In: Proceedings of the IEEE Conference on Computer Vision and Pattern Recognition (CVPR) (2016)

18. Simonyan, K., Zisserman, A.: Very deep convolutional networks for large-scale image recognition. arXiv preprint arXiv:1409.1556 (2015)
19. Zhou, Y., Sahak, H., Ba, J.: Using synthetic data for data augmentation to improve classification accuracy. In: Proceedings of the Workshop on Challenges in Deployable Machine Learning, ICML, Honolulu, Hawaii, USA (2023)
20. Tan, M., Le, Q.V.: EfficientNetV2: smaller models and faster training. In: Proceedings of the International Conference on Machine Learning (ICML) (2021)
21. Kingma, D.P., Welling, M.: Auto-encoding variational bayes. In: Proceedings of the 2nd International Conference on Learning Representations (ICLR), Banff, AB, Canada (2014)

Automated Multi-View Planning for Endovascular Aneurysm Repair Procedures

Baochang Zhang[1,2,4]($\boxtimes$), Yiwen Liu[1], Shuting Liu[1], Heribert Schunkert[2], Reza Ghotbi[3], and Nassir Navab[1]

[1] Computer Aided Medical Procedures, Technical University of Munich, Munich, Germany
[2] German Heart Center Munich, Munich, Germany
[3] German Centre for Cardiovascular Research, Munich Heart Alliance, Munich, Germany
[4] HELIOS Hospital west of Munich, Munich, Germany
baochang.zhang@tum.de

Abstract. During Endovascular aneurysm repair (EVAR) procedures, surgeons always require several views of vessel structures to accurately assess the size, shape, and location of the aneurysm, along with the surrounding vasculature. However, even expert surgeons often require multiple attempts to find a desired view, which leads to increased radiation exposure, high doses of contrast agents for patients, and time-consuming re-positioning of the C-arm. This paper introduces an automatic framework to provide optimal multi-view for the whole EVAR procedure. First, a 3D nnUNet is employed to extract geometric information and semantic information, providing accurate vascular and aneurysm segmentation as well as semantic bifurcation detection. Then, a semantic vessel tree model is built by integrating semantic information and geometric information. A local 3D plane at each critical bifurcation is fitted based on the centerlines surrounding this bifurcation, where we regard the estimated 3D local plane as a good view plane in patient physical space. Next, some 3D points are collected from these centerlines, projected onto the estimated local 3D plane, and transformed to the image domain to get the paired 2D points. Finally, based on the geometric information of the C-arm X-ray imaging device, the most informative view pose for C-arm positioning is solved via RANSAC Perspective-n-Point algorithm with the Levenberg-Marquardt optimization. Our work not only streamlines the surgical planning process, but also helps in customizing the patient-specific strategies to reduce risks and improve surgical outcomes. Our framework has been validated using an in-house dataset collected from 27 patients, which contains preoperative CTA data and intraoperative X-ray angiography images. The qualitative and quantitative results demonstrate the reliability and effectiveness of our approach. Meanwhile, our system achieved an average runtime of 6 min per patient.

Keywords: Multiple View Planning · Abdominal Aortic Aneurysm · EVAR procedures

---

B. Zhang and Y. Liu — These authors contributed equally to this work.

K. Drechsler et al. (Eds.): CLIP 2024, LNCS 15196, pp. 22–31, 2024.
https://doi.org/10.1007/978-3-031-73083-2_3

1   Introduction

In recent years, the rate of patients treated with Endovascular Aneurysm Repair (EVAR) procedures has increased notably. Typically, successful EVAR procedures demand the acquisition of multiple views to guide the intervention comprehensively. Each view serves a vital role in navigating the intricate vascular anatomy, ensuring precise placement of endovascular devices, and monitoring post-procedural outcomes. However, the safety of patients is still a major concern as imaging surveillance is required. This comes with risks associated with radiation exposure, contrast agent use, as well as increased costs.

With the exponential growth of medical imaging data and advancements in computational power, deep learning algorithms have demonstrated remarkable performance in various healthcare applications [19]. Utilizing preoperative CT angiography images, significant progress has been made in 3D vascular segmentation [3], aneurysm detection [17], aneurysm growth prediction [10], vascular centerline extraction [6], and vessel labeling [20] to assist EVAR procedures. Intra-operative X-ray images have also been the focus of numerous learning-based approaches to aid EVAR procedures, such as X-ray/CT registration [13] and 2D vessel segmentation [8]. However, few studies directly address the acquisition of the optimal surgeon's view. The definition of good views involves ensuring that the imaging provides clear, accurate, and comprehensive visualization of the relevant anatomical structures during the procedure. Fallavollita et al. [2] proposed a user interface concept enabling the surgeon to manually select the desired view before surgery, aiming to alleviate the challenges associated with constantly repositioning the angiographic C-arm during intervention. Tehlan et al. [16] suggested using an augmented reality head-mounted display for the surgeon to choose a desired X-ray view, providing corresponding C-arm configuration as visual feedback. Nevertheless, manually selecting these optimal views can be time-consuming and subjective, potentially leading to suboptimal outcomes and increased patient risks. Recently, Kausch et al. [9] introduced a convolutional neural network regression model to predict five degrees of freedom pose updates directly from the initial X-ray image in orthopedic surgery, facilitating automated C-arm positioning to achieve the desired view. However, this approach necessitates manual annotations of desired views for training, which is a labor-intensive and time-consuming task that significantly restricts its applicability.

In this paper, we introduce a complete framework for automatically providing multiple optimal views to guide EVAR procedure, thereby reducing the need for manual selection and minimizing procedural inefficiencies. The main contributions are as follows. (1) We leverage 3D nnUNet to extract geometric information and semantic information, where 3D center distance loss is proposed for accurate semantic bifurcation detection. (2) Our framework effectively integrates semantic geometric information extracted from patient-specific pre-operative CTA data and geometric information of C-arm X-ray imaging device for multi-view C-arm positioning. (3) We validate the feasibility and effectiveness of the proposed framework using CTA data of 27 patients and their corresponding intraoperative

X-ray angiography images, and the entire pipeline achieves an average runtime of 6 min per patient.

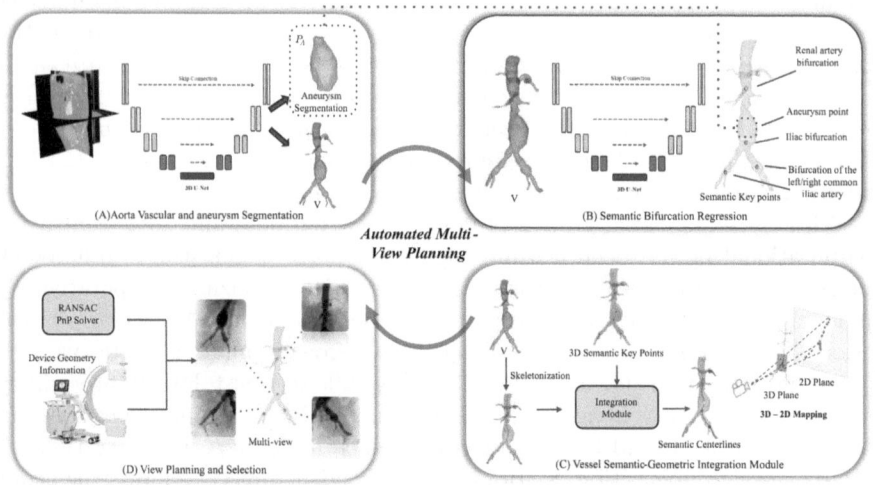

Fig. 1. An overview of the proposed pipeline for automated multi-view planning. The notation V represents the vascular segmentation, P_A is the aneurysm coordinates.

2   Method

2.1   Aorta and Aneurysm Segmentation

We employ a 3D nnU-Net [7] to segment both the aortic vascular and the aneurysms from pre-operative CTA data, trained with the Dice loss function, as shown in Fig. 1(A). The coordinates of the aneurysm P_A are further determined by calculating the centroid of the segmented aneurysm area.

2.2   Semantic Bifurcation Detection

This section focuses on extracting semantic information from pre-operative data. Relying solely on extracting and labeling the centerline from vascular segmentation V can lead to inaccuracies, particularly when the segmentation is not continuous [18]. To improve accuracy, we focus our semantic annotations exclusively on four key bifurcations. We utilize a 3D nnU-Net [7] to detect these bifurcations directly by regressing Gaussian heatmap kernels in a supervised learning manner, as illustrated in Fig 1(B).

For the ground truth, it has four channels, and the channel order represents distinct semantic information. Each channel $G \in R^{D \times H \times W}$ contains an unnormalized 3D Gaussian distribution centered on each key bifurcation:

$$G(i, j, k) = \exp\left(-\frac{(i - c_i)^2 + (j - c_j)^2 + (k - c_k)^2}{2\sigma^2}\right) \tag{1}$$

where (c_i, c_j, c_k) is the IJK coordinates of each bifurcation. Although heatmap-based regression method is commonly used in 2D key point detection [11], it faces challenges in 3D space due to spatial sparsity. To mitigate this, we set the σ value to 28mm to maximize non-overlapping area and minimize sparsity.

We downsampled the vascular segmentation results to help the network recognize global structures more easily and simplify learning, especially for distinguishing symmetric key points, like left common iliac artery bifurcation and right common iliac artery bifurcation. Additionally, we employ specialized loss functions to adapt the sparsity of 3D space.

Weighted Mean Squared Loss. To encourage the network to learn non-zero values, errors associated with non-zero values are given higher weight [4].

$$L_{\text{WMSE}} = \frac{1}{N} \sum_{i=1}^{N} w_i (y_i - \hat{y}_i)^2 \quad \text{where} \quad w_i = \begin{cases} 1.8 & \text{if } y_i > 0 \\ 1 & \text{otherwise} \end{cases} \tag{2}$$

Here, $\hat{y}_i$ and y_i represent the predicted and ground truth values. The weight w_i emphasize foreground values.

3D Center Distance Loss. While the weighted mean squared error loss L_{WMSE} can align predicted voxel values with the ground truth Gaussian distribution, it does not guarantee the accuracy of the predicted center point coordinates. Typically, the Gaussian distribution peaks at the coordinates of annotated bifurcations. While the Argmax function $torch.argmax()$ returns the peak coordinates of predicted Gaussian distributions, it is not differentiable. Inspired by the differentiable 2D SoftArgmax function [12], we implement a 3D variant that reduces the distance between the predicted and real center coordinates. It has been proved that the maximum value location can be approximated by a weighted sum of the predicted heatmap $G \in \mathbb{R}^{D \times H \times W}$ [5], namely taking the expectation of the probability map G. The predicted maximum value coordinate $\hat{c}$ is calculated as:

$$\hat{c} = SoftArgmax(G) = \sum_{n=1}^{3} \sum_{i=1}^{D} \sum_{j=1}^{H} \sum_{k=1}^{W} W_{n,i,j,k} G'_{i,j,k} \tag{3}$$

where the $G'_{i,j,k}$ is the softmax normalized value of G at location (i, j, k). With the location coordinate (i, j, k), we calculate weighted matrix $W \in \mathbb{R}^{3 \times D \times H \times W}$, which can be treated as 3D discrete normalized ramps along axis I, J, K. The notation n corresponds to these three channels:

$$W_{1,i,j,k} = \frac{i}{D}, W_{2,i,j,k} = \frac{j}{H}, W_{3,i,j,k} = \frac{k}{W} \tag{4}$$

Our center distance loss calculates the $L1$ Loss between the predicted and ground-truth key bifurcation:

$$L_{center} = |\hat{c} - c_{gt}| \tag{5}$$

Therefore, the composite loss uses coefficiency $\lambda = L_{WMSE}/L_{Center}$ to dynamically combine these two losses:

$$L_{total} = L_{WMSE} + \lambda L_{center} \tag{6}$$

2.3   Automated Optimal View Selection

Until now, the vascular segmentation V and the predicted bifurcation coordinates K are obtained. In order to integrate these geometric and semantic information together, a semantic vascular tree is built. For the centerline extraction from vascular segmentation V, the classic iterative thinning algorithm [14] is adopted. Then, based on topological analysis in 26 neighborhood system, all bifurcations, edges points and end points are identified from extracted centerline model. Predicted bifurcation coordinates K are adjusted to align with nearby bifurcations on the centerline model if they fall within a specified threshold. When we identify a nearest bifurcation point, we accurately determine the location of key points and provide relevant semantic information for those points. Next, we can fit a 3D plane based on the interested bifurcations and the surrounding branches. Of note, due to the multiple bifurcations exist above the renal artery, we apply an outlier exclusion algorithm to discard atypical vessel orientations to ensure that the planes accurately represented the majority of vessel orientations. For further details, please check Algorithm 1.

We then randomly sample some points on the centerline model around the interested key points. The 2D-3D point pairs are obtained by projecting these 3D points onto the fitted plane and transforming them into image domain, resulting in 2D IJ coordinates. RANSAC Perspective-n-Point (PnP) algorithm [1] is utilized to calculate the pose for virtual C-arm positioning. The goal of using the RANSAC algorithm is to identify and mitigate outlier effects to accurately estimate the object's pose. The intrinsic camera matrix provided to the algorithm is calculated based on the following geometric parameters of the C-arm device: the distance from the X-ray source to the isocenter of device is 742.5 mm; the distance from the detector to the isocenter of device is 517.15 mm; both the width and height of the detector are 432 mm, and the pixel size in detector is 0.3 mm. Meanwhile, to enhance stability across each sampling iterations, we apply the Levenberg-Marquardt algorithm [15] iteratively, calculating the reprojection error from the PnP solution. We then remove outlier points according to the error while ensuring a minimum number of points are maintained. Finally, we visualize our pose quality by rendering X-ray images using Digital Reconstructed Radiography (DRR) method [21].

Algorithm 1

Input: Vascular segmentation V, aneurysm position $P_A \in R^3$, predicted key bifurcations coordinates list $K = \{k_i \in R^3, i = 4\}$

Output: Final key bifurcations coordinate list $K_{final} = \{k_{final_i} \in R^3, i = 4\}$, adjacent points dict $K_{adj} = \{k_{final_i} : k_{adj_j} \in R^3, j = min(len(points), 90)\}$

1: Get vessel tree T and bifurcation points P_b from V via thinning algorithm
2: **for** k_i in K **do**
3:      Relocate k_i with nearest bifurcation point P_b within threshold $dist = 20mm$ as k_{final_i}, if larger than threshold, choose k_i as k_{final_i}
4:      Identify vessel sub-tree $T_{sub_k} = T_{k_{final_i}}$ from T associated with k_{final_i}. Sample adjacent points k_{adj_j} on T_{sub_k} for k_{final_i}.
5:      **if** k_{final_i} is not kidney or aneurysm key point or degree($k_{final_i} - T_{sub_k}$) = 3 **then**
6:          continue
7:      **else**
8:          **for** each branch t^* in T_{sub_k} **do**
9:              Let $S = T_{sub_k} \setminus \{t^*\}$ {Set S contains all branches except t^*}
10:              Compute the angle θ between the directional vector of t^* and the normal vector of the plane fitted to S
11:              **if** all angle is greater than a threshold=25 **then**
12:                  Exclude branch $t*$ from T_{sub_k}
13: Fit arbitrary local plane L using final key points K_{final} and adjacent points K_{adj}.

3   Experiments and Results

3.1   Dataset

An abdominal dataset is collected from 27 patients diagnosed with aneurysms. For each patient, it includes a preoperative CTA data obtained using a GE Revolution EVO CT scanner and some intraoperative X-ray angiography images. For CTA images, the reconstructed slice thickness ranges from $1mm$ to $3mm$ and in?plane spacing from $0.79mm$ to $1.34mm$. In addition, abdominal vascular mask, aneurysm mask and four key bifurcations from each CTA data are manually annotated using open source 3D-Slicer software. The four key bifurcations are the renal artery bifurcation, iliac bifurcation, and the bifurcations of the left and right common iliac arteries, as shown in Fig. 1(B).

3.2   Results on Vascular and Aneurysm Segmentation

The experiment results of vascular and aneurysm segmentation by five-fold cross validation are shown in Table 1. Obviously, the vascular and aneurysm segmentation show strong performance, laying a solid foundation for subsequent geometric analysis.

3.3   Results on Semantic Key Bifurcation Detection

The performance of semantic key bifurcation detection is evaluated by calculating the mean distance and variance between predicted coordinates and ground

Table 1. Experiment results on vascular and aneurysm segmentation by five-fold cross validation

	Dice	Precision	Recall	F1-Score	IoU
Vessel Segmentation	0.975	0.964	0.986	0.975	0.952
Aneurysm Segmentation	0.866	0.933	0.831	0.879	0.770

Table 2. Ablation study on semantic bifurcation detection evaluated by L1 distance error

L_{WMSE}	L_{center}	Input	Mean/mm ↓	Variance/mm^2 ↓
✓	✓	Segmentation	**10.968**	**8.954**
✓	✓	Raw CTA	22.313	165.515
✓		Segmentation	18.811	89.561
	✓	Segmentation	99.506	12057.061

truth coordinates of each key bifurcation. The performance of our method is shown in the first row of Table 2. Meanwhile, Table 2 also reflects outcomes from ablation studies testing various combinations of inputs and loss functions. Compared with taking raw CTA data as input, taking vascular segmentation as input has more advantages, resulting in less errors in the mean distance and distance criterions. While employing L_{WMSE} alone yields commendable regression outcomes, it is crucial to note that this loss function does not directly target bifurcation coordinates. Therefore, the integration of L_{center} loss can further enhances our results. Interestingly, when only L_{center} is used, it leads to the worst performance.

3.4 Results on View Planning

An ablation study is conducted on view pose solution, as shown in Table 3. The PnP re-projection error is employed as evaluation metric. From Table 3, traditional PnP algorithm fails in solving view pose. And our experiment results demonstrate that the RANSAC PnP algorithm [1] combined with the Levenberg-Marquardt (LM) algorithm [15] achieves very stable and accurate view pose solution.

Table 3. Ablation study on view pose solution evaluated by re-projection error

Algorithms	Mean /mm ↓	Variance/mm^2 ↓
Traditional PnP	3333.945	2.803×10^7
RANSAC PnP	1.060	0.060
RANSAC PnP + LM	**0.565**	**0.043**

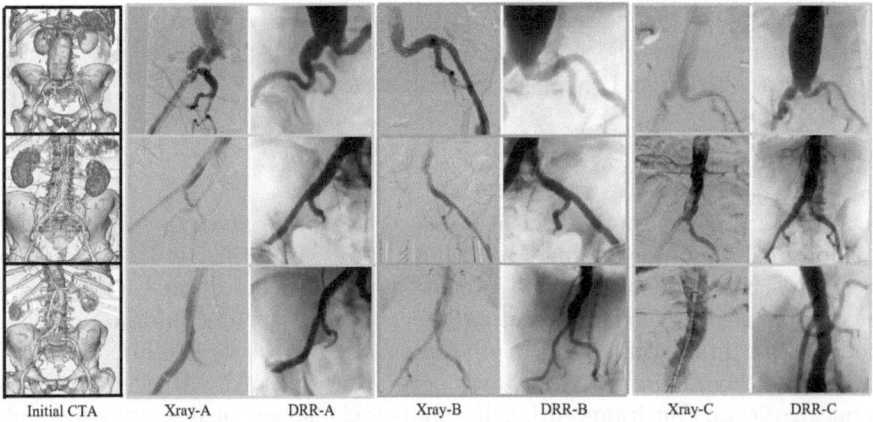

Initial CTA   Xray-A   DRR-A   Xray-B   DRR-B   Xray-C   DRR-C

Fig. 2. Visual comparison between real X-rays and DRRs generated from planned view poses in three patients. The locations of the comparison correspond to the 3D visualization of the CTA and are marked with different colors.

Since our work is the first to propose multi-view planning for EVAR procedures, there are no existing works available for direct comparison. To validate the effectiveness of our proposed framework, we compared our results with intra-operative X-Ray angiography images, as illustrated in Fig. 2. Compared to the X-ray images used by surgeons during intervention, the DRR images generated using our planned views are very similar to them and show clearer vascular projection anatomy. The surgeons further evaluated the planning views generated by our proposed framework in these 27 patients and they were confident that these planning views were sufficient to guide the surgery. Additionally, more planning views are shown in Fig. 3. Our proposed method can not only provide a view based on a single key point of interest, but also coordinate multiple key points

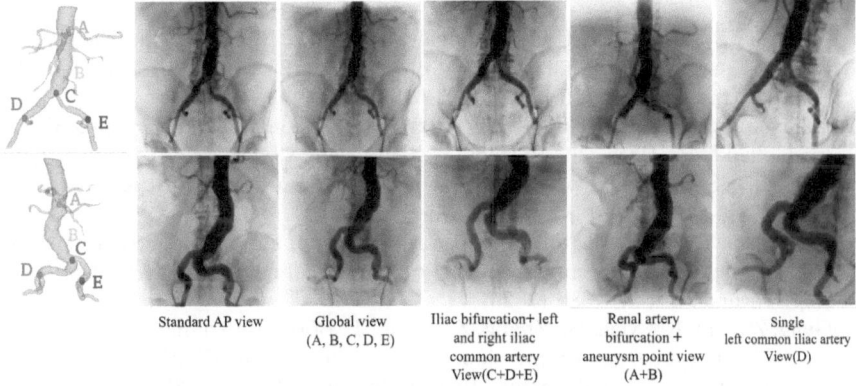

| Standard AP view | Global view (A, B, C, D, E) | Iliac bifurcation+ left and right iliac common artery View(C+D+E) | Renal artery bifurcation + aneurysm point view (A+B) | Single left common iliac artery View(D) |

Fig. 3. Different views based on various key points of interest.

of interest to plan the view. Compared to standard AP view, the planned views help surgeons understand complex vascular anatomy more easily and quickly, showing less overlap and clearer branching structures.

4 Conclusion

In this paper, we present an efficient framework that automatically provides multiple optimal views for guiding EVAR procedures. By utilizing 3D nnUNet for precise vascular segmentation and semantic bifurcation detection, we construct a semantic vessel tree model integrating geometric and semantic information. This model assists in identifying optimal viewing planes at critical bifurcations. We then accurately determine the C-arm pose using the RANSAC Perspective-n-Point algorithm. Our framework helps address challenges such as excessive radiation exposure, high contrast agent doses, and time-consumption repositioning, streamlining the surgical planning process and enabling patient-specific strategies. Validated with an in-house dataset from 27 patients, our system demonstrates reliability, effectiveness, and a practical runtime of 6 min per patient. Future work will focus on further refining the framework to enhance its adaptability to a wider range of vascular surgeries and integrating preoperative and intra-operative registration modules to smoothly apply preoperative view planning to intra-operative settings.

Acknowledgments. The project was supported by the Bavarian State Ministry of Science and Arts within the framework of the"Digitaler Herz-OP" project under the grant number 1530/891 02 and the China Scholarship Council (File No.202004910390). We also thank BrainLab AG for their partial support.

Disclosure of Interests. The authors have no competing interests to declare that are relevant to the content of this article.

References

1. Chum, O., Matas, J., Kittler, J.: Locally optimized RANSAC. In: Michaelis, B., Krell, G. (eds.) DAGM 2003. LNCS, vol. 2781, pp. 236–243. Springer, Heidelberg (2003). https://doi.org/10.1007/978-3-540-45243-0_31
2. Fallavollita, P., et al.: Desired-view controlled positioning of angiographic C-arms. In: Golland, P., Hata, N., Barillot, C., Hornegger, J., Howe, R. (eds.) MICCAI 2014. LNCS, vol. 8674, pp. 659–666. Springer, Cham (2014). https://doi.org/10.1007/978-3-319-10470-6_82
3. Fantazzini, A., et al.: 3D automatic segmentation of aortic computed tomography angiography combining multi-view 2D convolutional neural networks. Cardiovasc. Eng. Technol. **11**, 576–586 (2020)
4. Geng, Z., Sun, K., Xiao, B., Zhang, Z., Wang, J.: Bottom-up human pose estimation via disentangled keypoint regression. In: Proceedings of the IEEE/CVF Conference on Computer Vision and Pattern Recognition, pp. 14676–14686 (2021)
5. Goroshin, R., Mathieu, M.F., LeCun, Y.: Learning to linearize under uncertainty. Adv. Neural Inf. Process. Syst. **28** (2015)

6. He, J., et al.: Learning hybrid representations for automatic 3D vessel centerline extraction. In: Martel, A.L., et al. (eds.) MICCAI 2020. LNCS, vol. 12266, pp. 24–34. Springer, Cham (2020). https://doi.org/10.1007/978-3-030-59725-2_3
7. Isensee, F., Jaeger, P.F., Kohl, S.A., Petersen, J., Maier-Hein, K.H.: nnU-Net: a self-configuring method for deep learning-based biomedical image segmentation. Nat. Methods **18**(2), 203–211 (2021)
8. Kappe, K.O., Smorenburg, S.P., Hoksbergen, A.W., Wolterink, J.M., Yeung, K.K.: Deep learning-based intraoperative stent graft segmentation on completion digital subtraction angiography during endovascular aneurysm repair. J. Endovasc. Ther. **30**(6), 822–827 (2023)
9. Kausch, L., et al.: Toward automatic C-arm positioning for standard projections in orthopedic surgery. Int. J. Comput. Assist. Radiol. Surg. **15**, 1095–1105 (2020)
10. Kim, S., et al.: Deep learning on multiphysical features and hemodynamic modeling for abdominal aortic aneurysm growth prediction. IEEE Trans. Med. Imag. **42**(1), 196–208 (2022)
11. Li, J., Su, W., Wang, Z.: Simple pose: rethinking and improving a bottom-up approach for multi-person pose estimation. In: Proceedings of the AAAI Conference on Artificial Intelligence, vol. 34, pp. 11354–11361 (2020)
12. Luvizon, D.C., Tabia, H., Picard, D.: Human pose regression by combining indirect part detection and contextual information. Comput. Graph. **85**, 15–22 (2019)
13. Meng, C., Wang, Q., Guan, S., Sun, K., Liu, B.: 2d–3d registration with weighted local mutual information in vascular interventions. IEEE Access **7**, 162629–162638 (2019)
14. Palágyi, K., et al.: A sequential 3D thinning algorithm and its medical applications. In: Insana, M.F., Leahy, R.M. (eds.) IPMI 2001. LNCS, vol. 2082, pp. 409–415. Springer, Heidelberg (2001). https://doi.org/10.1007/3-540-45729-1_42
15. Ranganathan, A.: The Levenberg-Marquardt algorithm. Tutoral LM Algorithm **11**(1), 101–110 (2004)
16. Tehlan, K., Winkler, A., Roth, D., Navab, N.: X-ray device positioning with augmented reality visual feedback. In: 2022 IEEE Conference on Virtual Reality and 3D User Interfaces Abstracts and Workshops (VRW), pp. 870–871. IEEE (2022)
17. Timmins, K.M., Van der Schaaf, I.C., Vos, I.N., Ruigrok, Y.M., Velthuis, B.K., Kuijf, H.J.: Geometric deep learning using vascular surface meshes for modality-independent unruptured intracranial aneurysm detection. IEEE Trans. Med. Imag. **42**(11), 3451–3460 (2023)
18. Virga, S., Dogeanu, V., Fallavollita, P., Ghotbi, R., Navab, N., Demirci, S.: Optimal C-arm positioning for aortic interventions. In: Handels, H., Deserno, T.M., Meinzer, H.P., Tolxdorff, T. (eds.) Bildverarbeitung für die Medizin 2015. I, pp. 53–58. Springer, Heidelberg (2015). https://doi.org/10.1007/978-3-662-46224-9_11
19. Wang, Y., et al.: Deep learning model for predicting the outcome of endovascular abdominal aortic aneurysm repair. Indian J. Surg. **85**(Suppl 1), 288–296 (2023)
20. Yao, L., et al.: TaG-Net: topology-aware graph network for centerline-based vessel labeling. IEEE Trans. Med. Imag. 42(11), 3155–3166 (2023)
21. Zhang, B., et al.: A patient-specific self-supervised model for automatic X-ray/CT registration. In: Greenspan, H., et al. (eds.) Medical Image Computing and Computer Assisted Intervention – MICCAI 2023. MICCAI 2023. LNCS, vol. 14228. Springer, Cham (2023). https://doi.org/10.1007/978-3-031-43996-4_49

HTSeg: Hybrid Two-Stage Segmentation Framework for Intestine Segmentation from CT Volumes

Qin An[1]([✉]), Hirohisa Oda[2], Yuichiro Hayashi[1], Takayuki Kitasaka[3], Aitaro Takimoto[4], Akinari Hinoki[4], Hiroo Uchida[4], Kojiro Suzuki[5], Masahiro Oda[1,6], and Kesaku Mori[1,6,7]

[1] Graduate School of Informatics, Nagoya University, Nagoya, Japan
anqin1117@gmail.com
[2] School of Management and Information, University of Shizuoka, Shizuoka, Japan
[3] School of Information Science, Aichi Institute of Technology, Toyota, Japan
[4] Graduate School of Medicine, Nagoya University, Nagoya, Japan
[5] Department of Radiology, Aichi Medical University, Nagakute, Japan
[6] Information Technology Center, Nagoya University, Nagoya, Japan
[7] Research Center for Medical Bigdata, National Institute of Informatics, Tokyo, Japan

Abstract. This paper proposes a semi-supervised intestine segmentation method from CT volumes. Our method can use densely and sparsely annotated CT volumes for training to reduce the labor of manually annotating intestines. The proposed Hybrid Two-stage Segmentation (HTSeg) framework consists of two networks, a 2D swin-transformer-based network as the first stage and a 3D network as the second stage. In the first stage, we use 6964 labeled CT slices to train the 2D Swin U-Net. The trained 2D Swin U-Net is used to generate pseudo-labels for sparse annotation data. In the second stage, we use sparsely annotated datasets with pseudo-labels and densely annotated datasets to train a 3D multi-view symmetrical network (MVSNet). Experimental results showed that the Dice score of the proposed method was 74.70%, which was 1.03% higher than just using MVSNet. Compared with the other four previous methods (3D U-Net, CPS, EM, MT), the proposed method produced competitive segmentation performance. The code can be found at: https://github.com/MoriLabNU/semi-pseudo-labels.

Keywords: Intestine segmentation · Semi-supervision · Pseudo-label · Sparse annotation

1 Introduction

This paper proposes a semi-supervised intestine segmentation method for CT volumes. Given the complexity of the intestine, there is a desire to utilize CT volumes that are not fully annotated. Our method can incorporate both densely and sparsely annotated CT volumes for training.

© The Author(s), under exclusive license to Springer Nature Switzerland AG 2024
K. Drechsler et al. (Eds.): CLIP 2024, LNCS 15196, pp. 32–41, 2024.
https://doi.org/10.1007/978-3-031-73083-2_4

The intestine is a long organ in the human body, highly folded in the abdominal cavity and surrounded by various organs with complex structures. Intestinal obstruction [1–4] is a medical condition characterized by blockage of the intestines. Intestine segmentation can help clinicians know the structure of the intestine and locate the position of intestinal obstruction from CT volumes.

Full-supervised learning-based methods have been investigated for many years, with most methods gradually being applied in medical image segmentation [5–8]. Specifically, U-Net [9] and its variants [10–12] have achieved good results in various organ segmentation tasks. However, these methods require extensive voxel-level annotations, which is especially challenging for CT volumes composed of hundreds of CT slices. Additionally, correct annotations of CT volumes require expertise from medical professionals. Exploring semi-supervised learning segmentation methods that leverage unlabeled data is useful in reducing the amount of annotations required.

Non-machine-learning-based methods, such as intensity thresholding and region growing, have been explored for intestine segmentation. Machine learning has also been used in several works [13–18]. However, the intestine has similar intensities to the surrounding organs on CT volumes and has a complex structure. Furthermore, the shape and location of the intestines vary from person to person. These methods do not comprehensively describe the information about the intestines, which results in the segmentation effect failing to meet expectations.

Considering these challenges, we propose the hybrid two-stage segmentation (HTSeg) framework to enhance the accuracy of segmenting intestines in CT volumes. Our method aims to segment intestine regions and assist clinicians in quickly and accurately diagnosing intestinal diseases. The contributions of the proposed method are summarized as follows:

- **New segmentation method for semi-supervised learning.** This method is a two-stage network that combines a 2D transformer and a 3D convolutional neural network (CNN). In the first stage, we employ large-scale labeled CT slices to train a 2D Swin U-Net [19] for generating pseudo-labels [20]. In the second stage, we concatenate CT patches from densely and sparsely labeled data and corresponding predictions from 2D Swin U-Net. These are used as inputs to train a 3D multi-view symmetrical network (MVSNet) [21] as the final model.
- **Combination of pseudo-labels and sparsely-annotated labels.** The pseudo-labels are generated by the 2D Swin U-Net and sparse annotation is labeled by clinicians to create semi-pseudo-labels for sparse label data. This approach helps reduce the unreliability of the pseudo-labels that just rely on the model's predictions.

2   Method

2.1   Overview

Our HTSNet is a two-stage framework that combines the 2D transformer and 3D CNN architecture. It uses different dimensional features to reduce insufficient

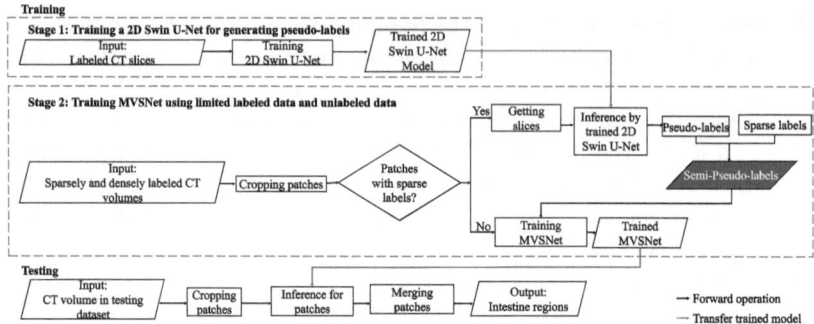

Fig. 1. Flowchart of the proposed method. In the first stage, we use labeled slices to train a 2D Swin U-Net. The trained model is used to generate pseudo-labels. In the second stage, the densely labeled and sparsely labeled data are used to train MVSNet. Finally, we employ the trained MVSNet to infer the testing dataset. The content in the green box is the main contribution point in the paper. (Color figure online)

intestine segmentation from CT volumes. Our proposed method is based on semi-supervised learning, where the first stage aims to train a model to generate pseudo-labels for sparsely labeled data. The second stage utilizes limited labeled data and a large amount of sparse labeled data to train a 3D CNN network.

Different from common-used pseudo-labels, we use semi-pseudo-labels that fuse sparse labels and pseudo-labels for sparsely labeled data. This approach aims to reduce the influence of incorrect pseudo-labels that may result from mistakes made by the model during the labeling process. These mistakes can be used to further train the model, potentially reinforcing and amplifying training errors. The semi-pseudo-labels can decrease the influence of incorrect pseudo-labels by utilizing sparse labels created by clinicians.

Our dataset includes limited dense-labeled CT volumes and plenty of sparse-labeled CT volumes. We obtain labeled axial, sagittal, and coronal CT slices from dense-labeled CT volumes to train the 2D Swin U-Net. The trained model is used to generate pseudo-labels for CT volumes with sparse labels. Then we use the semi-pseudo-label that combines the pseudo-label and sparse label to train MVSNet as the final segmentation model. The flowchart of the overall method is shown in Fig. 1.

2.2   Semi-pseudo-labels

In this work, pseudo-labeling relies on the prediction of the first model, which can propagate and amplify errors from the model's predictions. These errors can lead to incorrect training of the second network. Therefore, we propose the 'semi-pseudo-label', combining the sparse-labels and the pseudo-labels to increase the correction of the pseudo-labels Fig. 2.

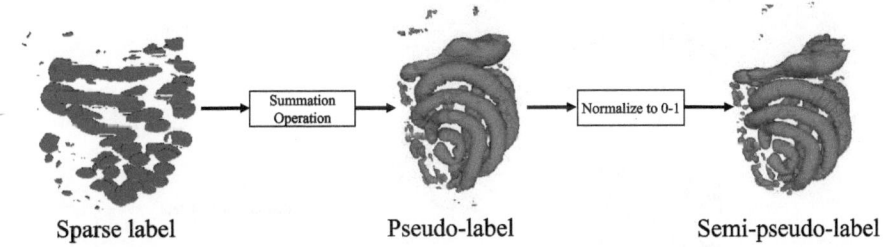

Fig. 2. The process of generating semi-pseudo-labels for sparsely-labeled data. We obtain the sum of the sparse- and pseudo-labels and normalize the result to 0, 1. The semi-pseudo-label consists of values $\{0, 1\}$.

2.3 Hybrid Two-Stage Segmentation (HTSeg) Framework

In this research, we utilize a limited number of densely and plenty number of sparsely labeled data. Using only a limited amount of densely labeled data to train the network could lead to overfitting and poor segmentation performance. Similarly, sparsely labeled data may not provide enough information for the model to learn complex features in the data, limiting its ability to make accurate predictions. To address these challenges and fully utilize the small amount of densely labeled data while better leveraging sparsely labeled data, we propose a two-stage model called the hybrid two-stage segmentation (HTSeg) framework.

The CT volumes, as three-dimensional images, consist of a series of consecutive two-dimensional CT slices, which can be obtained from axial, sagittal, and coronal planes, respectively. Based on this characteristic, we can get thousands of CT slices from the densely-labeled CT volumes. In the first stage, we use these thousands of CT slices to train a 2D Swin U-Net. The 2D Swin U-Net is used to generate pseudo-labels for sparsely-labeled data. However, since the structure of intestines in CT volumes is complex, both the inter-slices and intra-slices features are crucial. Moreover, in the first stage, we only utilize 2D CT slices leading to ignore the inter-slice features. This limitation may result in the underperformance of the network due to insufficient semantic features.

To address this issue, in the second stage, we train a 3D convolutional neural network (CNN), MVSNet, using densely labeled and sparsely labeled CT volumes. Different from using only the original data, we concatenate the original data and the corresponding prediction of 2D Swin U-Net as inputs of the MVSNet. In this stage, we can use hybrid features [22] from 2D and 3D to train MVSNet. The structure of the proposed method is shown in Fig. 3.

2.4 Loss Function

The method involves training two networks, each with a distinct loss function. The 2D Swin U-Net is trained using a supervised loss function, while the MVSNet is trained using both supervised and unsupervised loss functions. The overall

Step 1: Training 2D Swin U-Net

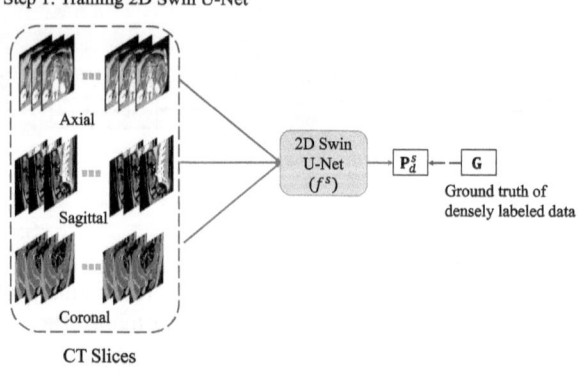

Step 2: Training MVSNet

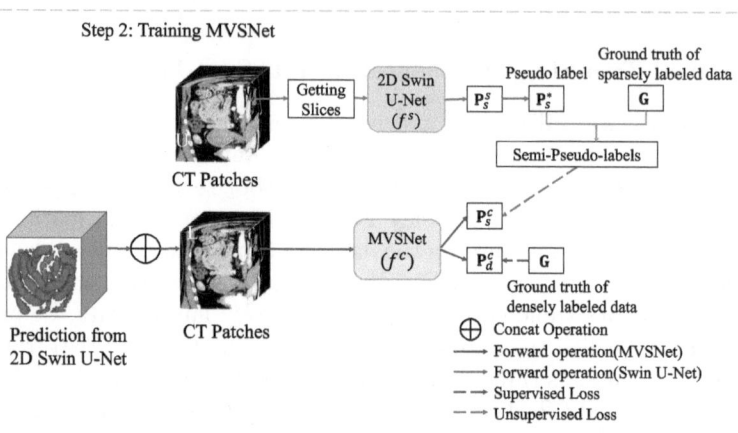

Fig. 3. The structure of the proposed method. In the first stage, CT slices from densely labeled data train a 2D Swin U-Net. In the second stage, original CT patches from both sparsely and densely labeled data are concatenated with the corresponding prediction of the 2D Swin U-Net as input to train the MVSNet.

loss function includes the supervised and unsupervised loss functions and is calculated by the equation $L_{overall} = L_{sup} + L_{un}$.

In the first stage, the main task is to train the 2D Swin U-Net for generating pseudo-labels. We use the CT slices with their corresponding ground truth as inputs, so only the supervised loss function is used. We use a combination of cross-entropy (CE) loss L_{ce} and Dice loss L_{dice} as the supervised loss function [23]. The CE loss focuses on optimizing the pixel-wise classification accuracy, while the Dice loss emphasizes the spatial overlap between the predicted and ground truth masks. Combining these two losses can improve segmentation performance and produce more accurate segmentation.

$$L_{sup}(\mathbf{P}_d^s, \mathbf{G}) = \alpha L_{ce}(\mathbf{P}_d^s, \mathbf{G}) + (1 - \alpha)L_{dice}(\mathbf{P}_d^s, \mathbf{G}), \tag{1}$$

where $\mathbf{P}_d^s$ represent the predictions of the 2D network, $\mathbf{G}$ represents the ground truth.

In the second stage, we use densely labeled and sparsely labeled data, both supervised and unsupervised loss functions are used. We just use Dice loss L_{dice} as the unsupervised loss function.

$$I_{sup}(\mathbf{P}_d^c, \mathbf{G}) = \alpha L_{ce}(\mathbf{P}_d^c, \mathbf{G}) + (1 - \alpha)L_{dice}(\mathbf{P}_d^c, \mathbf{G}), \qquad (2)$$

$$L_{un}(\mathbf{P}_s^c, \mathbf{P}_s^*) = L_{dice}(\mathbf{P}_s^c, \mathbf{P}_s^*), \qquad (3)$$

where $\mathbf{P}_d^c$ and $\mathbf{P}_s^c$ represent the prediction of MVSNet for densely labeled and sparsely labeled data, respectively. $\mathbf{P}_s^*$ represents $\mathbf{P}_s^c$'s pseudo-labels generated in first stage. Here, we set the weight $\alpha = 0.5$.

3   Experimental Results

3.1   Dataset and Experimental Settings

Our experiments used 171 CT volumes from various patients, comprising 13 densely labeled and 158 sparsely labeled datasets. Densely labeled CT volumes indicate that clinicians have labeled intestines in hundreds of consecutive slices, while sparsely labeled CT volumes indicate labeling in only some of the non-consecutive slices. Before training, we performed an interpolation operation on the original CT images. The CT volume specification is shown in Table 1.

The program code was implemented in PyTorch and executed on an NVIDIA A100 GPU. The model was trained for up to thousands of iterations, with early stopping implemented applied if the best Dice score on validation remained unchanged for 200 iterations. The training used the SGD optimizer with a poly-learning rate strategy, starting with an initial learning rate of 0.01.

Table 1. CT volumes before and after interpolation. We present detailed information on the dataset.

	Original	Interpolation result
Slice size (pixels)	512×512	(281 - 463)×(281 - 463)
Number of slices	198 - 546	396 - 762
Resolution (mm^3)	$(0.549-0.904)\times$ $(0.549-0.904)\times(1.000-2.000)$	1.000×1.000×1.000

3.2   Results

Table 2 shows the results of various methods including classical full-supervised learning and semi-supervised learning methods. We evaluated these methods based on Dice score, recall, and precision rates. Additionally, we calculated the standard deviation (SD) value based on all test cases.

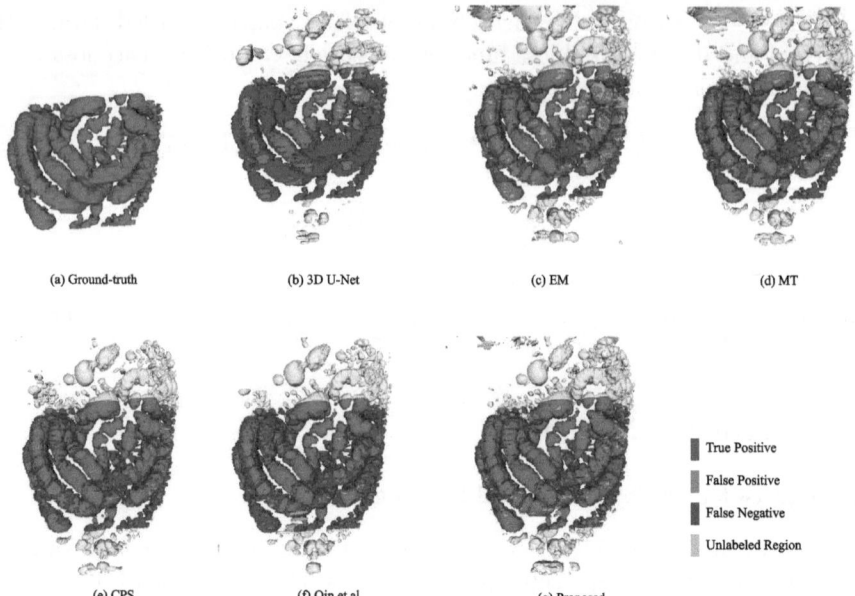

(a) Ground-truth   (b) 3D U-Net   (c) EM   (d) MT

(e) CPS   (f) Qin et al.   (g) Proposed

■ True Positive
■ False Positive
■ False Negative
■ Unlabeled Region

Fig. 4. Intestine segmentation results from different methods. (a) is the ground truth; (b) is the result of 3D U-Net; (c)-(f) are the results of previous semi-supervised learning methods; (g) is the result of the proposed method. We use different colors to present the different segmentation regions.

The classical full-supervised learning method 3D U-Net achieved a 39.66% Dice score. In comparison, CPS [23], EM [24], MT [25], Qin et al. [21], and our proposed method, which utilize both densely and sparsely labeled data achieved a 73.11%, 68.97%, 68.30%, 73.67%, and 74.70% Dice score, respectively. The performance of our method was better than the other previous method. The intestine segmentation results are shown in Fig. 4.

4   Discussions

We propose a new two-stage framework to segment the intestine from CT volumes. Our framework combines transformer and CNN architecture and introduces the concept of semi-pseudo-labels. Compared to the classical full-supervised learning method 3D U-Net, which is trained only on limited densely labeled data, our proposed method achieves competitive performance.

From Table 2, it is evident that the semi-supervised learning method is significantly more effective than the fully supervised learning method. From Fig 4, we can see that our method segments more intestine regions than other methods. We deduce that a small amount of densely labeled data is insufficient to fully leverage fully supervised learning methods, making semi-supervised learning a preferable option for tasks lacking abundant densely labeled data. Com-

Table 2. Evaluation results of different methods. We show the results by percentage and the bold font presents the best performance of each evaluation term.

Methods	Dice (%)	Recall (%)	Precision (%)
3D U-Nett [12]	39.66±15.67	81.88±12.71	28.72±13.73
CPS [23]	73.11±9.49	**85.23±10.90**	67.26±11.45
EM [24]	68.97±9.70	82.98±10.94	62.07±11.13
MT [25]	68.30±12.71	82.94±12.55	60.90±13.36
Qin et al. [21]	73.67±8.41	84.10±9.89	67.55±8.30
Proposed	**74.70±8.86**	83.72±10.84	**70.41±10.69**

pared to other semi-supervised learning methods, our proposed method not only uses pseudo-labels but also creates semi-pseudo-labels that retain the accuracy of manual labels based on pseudo-labels and complement the missing parts of sparse labels, ultimately enhancing network performance.

We also observe that the intestines can overlap with surrounding organs of similar intensity, which increases the difficulty of segmentation and introduces more uncertainty in the pseudo-labeling of the boundary. To address this challenge, we can focus on refining the boundary part to enhance the reliability of the pseudo-labels, thereby improving the model's performance.

5   Conclusions

This paper proposes the HTSeg framework for intestine segmentation from CT volumes to help clinicians effectively diagnose intestine diseases. The HTSeg framework, including a 2D Swin U-Net and an MVSNet, performs competitive segmentation from the densely and sparsely labeled data. Additionally, the method creates the semi-pseudo-label, which reduces the unreliable that pseudo-labels just rely on the prediction of the network. The performance of the proposed method in another dataset should be further explored. Furthermore, as mentioned in the discussion, segmentation of the boundary faces more challenges than other parts. Our method has a space for improvement by focusing more on refining the boundary regions to enhance accuracy.

Acknowledgments. Thanks for the help and advice from Mori laboratory. A part of this research was supported by the Hori Sciences and Arts Foundation, MEXT/JSPS KAKENHI (24H00720, 22H03203), the JSPS Bilateral International Collaboration Grants, and the JST CREST (JPMJCR20D5).

References

1. Smith DA, Kashyap S, N.S.: Bowel obstruction. StatPearls (1 Aug 2022)
2. Sinicrope, F.: Ileus and Bowel Obstruction. Holland-Frei Cancer Medicine. 6th edition. Hamilton BC Decker (2003)
3. Bower, K.L., Lollar, D.I., Williams, S.L., Adkins, F.C., Luyimbazi, D.T., Bower, C.E.: Small bowel obstruction. Surg. Clin. **98**(5), 945–971 (2018)
4. Bogusevicius, A., Pundzius, J., Maleckas, A., Vilkauskas, L.: Computer-aided diagnosis of the character of bowel obstruction. Int. Surg. **84**(3), 225–228 (1999). http://europepmc.org/abstract/MED/10533781
5. Roth, H.R., et al.: Deep learning and its application to medical image segmentation. Med. Imaging Technol. **36**(2), 63–71 (2018)
6. Zhou, Z., Rahman Siddiquee, M.M., Tajbakhsh, N., Liang, J.: UNet++: a nested U-net architecture for medical image segmentation. In: Deep Learning in Medical Image Analysis and Multimodal Learning for Clinical Decision Support: 4th International Workshop, DLMIA 2018, and 8th International Workshop, ML-CDS 2018, Held in Conjunction with MICCAI 2018, Granada, Spain, September 20, 2018, Proceedings 4, pp. 3–11. Springer (2018). https://doi.org/10.1007/978-3-030-00889-5_1
7. Ramesh, K., Kumar, G.K., Swapna, K., Datta, D., Rajest, S.S.: A review of medical image segmentation algorithms. EAI Endorsed Trans. Pervasive Health Technol. **7**(27), e6–e6 (2021)
8. Zeng, Y., Tsui, P.H., Wu, W., Zhou, Z., Wu, S.: Fetal ultrasound image segmentation for automatic head circumference biometry using deeply supervised attention-gated V-Net. J. Digit. Imaging **34**(1), 134–148 (2021)
9. Ronneberger, O., Fischer, P., Brox, T.: U-Net: convolutional networks for biomedical image segmentation. In: International Conference on Medical Image Computing and Computer-assisted Intervention, LNCS 9351. pp. 234–241. Springer (2015). https://doi.org/10.1007/978-3-319-24574-4_28
10. Xiao, X., Lian, S., Luo, Z., Li, S.: Weighted res-Unet for high-quality retina vessel segmentation. In: 2018 9th International Conference on Information Technology in Medicine and Education (ITME), pp. 327–331. IEEE (2018)
11. Cai, S., Tian, Y., Lui, H., Zeng, H., Wu, Y., Chen, G.: Dense-Unet: a novel multiphoton in vivo cellular image segmentation model based on a convolutional neural network. Quant. Imaging Med. Surg. **10**(6), 1275 (2020)
12. Çiçek, Ö., Abdulkadir, A., Lienkamp, S.S., Brox, T., Ronneberger, O.: 3D U-Net: learning dense volumetric segmentation from sparse annotation. In: International Conference on Medical Image Computing and Computer-assisted Intervention, LNCS 9901, pp. 424–432. Springer (2016). https://doi.org/10.1007/978-3-319-46723-8_49
13. Zhang, W., Kim, H.M.: Fully automatic colon segmentation in computed tomography colonography. In: 2016 IEEE International Conference on Signal and Image Processing (ICSIP), pp. 51–55. IEEE (2016)
14. Barr, K., Laframboise, J., Ungi, T., Hookey, L., Fichtinger, G.: Automated segmentation of computed tomography colonography images using a 3D U-Net. In: SPIE Medical Imaging 2020: Image-Guided Procedures, Robotic Interventions, and Modeling, vol. 11315, pp. 635–641 (2020)
15. Bert, A., et al.: An automatic method for colon segmentation in CT colonography. Comput. Med. Imaging Graph. **33**(4), 325–331 (2009). https://doi.org/10.1016/j.compmedimag.2009.02.004

16. Sato, Y., et al.: Tissue classification based on 3D local intensity structures for volume rendering. IEEE Trans. Visual Comput. Graphics **6**(2), 160–180 (2000)

17. Frimmel, H., Näppi, J., Yoshida, H.: Centerline-based colon segmentation for CT colonography. Med. Phys. **32**(8), 2665–2672 (2005)

18. Rajamani, K., et al.: Segmentation of colon and removal of opacified fluid for virtual colonoscopy. Pattern Anal. Appl. **21**(1), 205–219 (2018)

19. Cao, H., et al.: Swin-Unet: Unet-like pure transformer for medical image segmentation. In: European conference on computer vision, pp. 205–218. Springer (2022). https://doi.org/10.1007/978-3-031-25066-8_9

20. Zou, Y., Zhang, Z., Zhang, H., Li, C.L., Bian, X., Huang, J.B., Pfister, T.: Pseudoseg: Designing pseudo labels for semantic segmentation. arXiv preprint arXiv:2010.09713 (2020)

21. Qin, A., et al.: Intestine Segmentation from CT Volume based on Bidirectional Teaching. In: SPIE Medical Imaging 2024: Image Processing (accepted), vol. 12926, pp. 238–243 (2024)

22. Li, X., Chen, H., Qi, X., Dou, Q., Fu, C.W., Heng, P.A.: H-Denseunet: hybrid densely connected UNet for livor and tumor segmentation from CT volumes. IEEE Trans. Med. Imaging **37**(12), 2663–2674 (2018)

23. Chen, X., Yuan, Y., Zeng, G., Wang, J.: Semi-supervised semantic segmentation with cross pseudo supervision. In: Proceedings of the IEEE/CVF Conference on Computer Vision and Pattern Recognition, pp. 2613–2622 (2021)

24. Vu, T.H., Jain, H., Bucher, M., Cord, M., Pérez, P.: ADVENT: adversarial entropy minimization for domain adaptation in semantic segmentation. In: Proceedings of the IEEE/CVF Conference on Computer Vision and Pattern Recognition, pp. 2517–2526 (2019)

25. Tarvainen, A., Valpola, H.: Mean teachers are better role models: weight-averaged consistency targets improve semi-supervised deep learning results. Adv. Neural Inf. Proc. Syst. **30** (2017)

EnPrO: Enhancing Precision Through Optimization in Image-Guided Spine Surgical Procedures

Srivibha Parthasarathy[1], Durga R[1(✉)], Venkateswaran N[2], Vivek Maik[1],
Aparna Purayath[1], Manojkumar Lakshmanan[1],
and Mohanasankar Sivaprakasam[1]

[1] Healthcare Technology Innovation Centre (HTIC), Indian Institute of Technology
Madras (IITM), Chennai, India
`{durga.r,maik.vivek,aparna_p}@htic.iitm.ac.in`
[2] Department of Biomedical Engineering, Sri Sivasubramaniya Nadar College of
Engineering (SSN), Chennai, India

Abstract. Advancements in intra-operative visualization have accelerated the adoption of C-Arm Fluoroscopy imaging modalities within Image-Guided Spine Surgery (IGSS) procedures. The proposed research provides a novel technique for improving precision in IGSS via EnPrO by refining the mapping of 2D fluoroscopic images to the patient's anatomy. The fundamental strategy is to minimize reprojection error (RPE) by picking optimal fiducial sites via weighted norm approximation. In EnPrO, we propose two methods to perform optimization for fiducial weights, namely, using fiducial coordinates and using a camera projection matrix (CPM). Using EnPrO, the C-Arm imaging distortion that contributes to increasing RPE can be identified and excluded from the IGSS calibration procedure. Using EnPrO, fiducials were chosen based on weights obtained using the fiducial coordinates and the CPM clearly showed an average RPE decrease of 6.96% and 8.36% respectively. The implementation of the project can be found in this repository: **EnPrO - GitHub**

Keywords: Image-Guided Spine Surgery · Fiducial Localization · Reprojection Error · Camera Calibration · Weighted Norm Optimization

1 Introduction

Recent advancements in IGSS enable accurate localization of unseen anatomy such as spinal structures when operated in Minimally Invasive Surgery (MIS) mode [7]. This necessitates the need to improve intra-operative visualization using various imaging modalities such as CT, X-ray, fluoroscopy MRI, etc. [3]. This paper focuses on the C-Arm fluoroscopy modality which is brought into IGSS using a calibration drum allowing for precise measurement between the C-Arm fluoroscopic image and the patient anatomy. Calibration drums usually

K. Drechsler et al. (Eds.): CLIP 2024, LNCS 15196, pp. 42–52, 2024.
https://doi.org/10.1007/978-3-031-73083-2_5

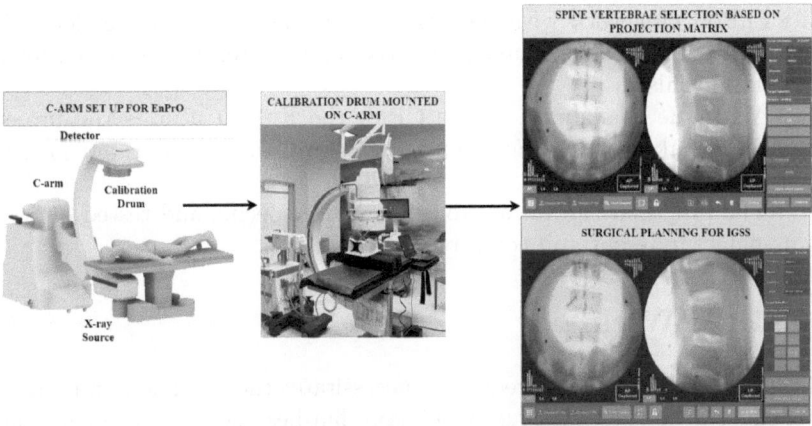

Fig. 1. Schematic of the Spine C-Arm image acquisition with fiducials overlayed using the calibration drum and shots captured in anterior-posterior (AP) and lateral-posterior (LP) views.

consist of embedded metallic markers and fiducials whose detection forms one of the primary requirements in IGSS [6]. Through the detection of the fiducials, all the algorithms of the IGSS pipeline such as distortion correction, point detection, calibration, etc. could be carried out which will enable the IGSS procedure [11]. The precision of the IGSS is very much dependent on the efficacy of the above computation algorithms which in turn depend on the fiducials detection [14].

Several factors can impede accurate fiducial recognition, including C-Arm distortion, fiducial occlusion owing to surgical instrument placement or diminished contrast in the C-Arm machine, flaws in the imaging method, and so on [13]. The twofold possibility from this drawback is that we detect the fiducials with severe distortion to hamper the RPE or we dont detect the fiducials which have minimum distortion. In either case, it is imperative to decide to either remove the distorted fiducial from the computational process leading the better precision or redo the image capture again such as to detect the fiducials with minimum distortion. Since the clinically acceptable error is <2mm, even a meagre reduction in the RPE proves to improve the accuracy of the process greatly [8]. During clinical data testing, it was observed that the accuracy of the IGSS improved significantly when some fiducials were excluded during calibration which meant that certain fiducials could contribute substantially to the RPE than others. In such instances, removing these fiducials may reduce the RPE. Therefore, selecting the optimal fiducials to achieve the lowest RPE leading to the highest precision is essential.

The main contributions of the proposed EnPrO are:

(i) EnPrO successfully identifies the fiducials that contribute the most and least to the precision of the IGSS.

(ii) EnPrO can work with only the fiducial positions independent of the CPM which is useful as the camera matrix cannot often be generated for high distortion fiducials.

(iii) With the availability of the CPM, the EnPrO utilizes the projected points which makes the optimization more robust on the criteria of producing the least RPE.

(iv) The proposed EnPrO approach is device-specific and tested on two C-Arm manufacturers but still could be made agnostic.

2   Related Works

Medical imaging computing frequently necessitates the detection and optimization of small structures, which provide major hurdles due to issues such as intensity changes, indistinct boundaries, and aberrations in images. Also, anatomical information must be kept intact because any anatomical deviation during computation could result in incorrect surgical interpretations. To address these challenges, many researchers have proposed innovative algorithms, Zhang et al. [13] pioneered a method for automatic detection and removal of fiducial projections in C-Arm calibration images and Bertelsen et al. [1] presented a novel automatic method for C-Arm distortion correction and calibration, emphasizing the importance of fiducial identification in enhancing the accuracy of IGSS. Their approach aimed to improve precision and reliability by utilizing fiducials for distortion correction and calibration processes, thus contributing to more accurate surgical interventions. D. H. Tungadio et al. [12] introduced a weighted and iterative weighted least squares method for power estimation, overcoming the convergence issues in linear programming methods and the constraints handling limitations in Newton and Gradient methods. El Hazzat et al. [2] and F. Mai and Y. S. Hung [9] both focused on improving 3D reconstruction accuracy by minimizing RPE. El Hazzat et al. used a binocular stereo-vision system with local bundle adjustment and the Levenberg-Marquardt algorithm, while Mai and Hung employed a factorization-based algorithm and the augmented Lagrangian method for projective reconstruction.

3   Methodology

For the proposed method, we utilize an in-house designed calibration drum which has 81 fiducials - 17 calibration fiducials and 64 distortion fiducials. To accurately map the patient space with the image space, the non-coplanarity is maintained in the distribution of the fiducials. The inputs to the EnPrO would be these detected distortion and calibration fiducials which would be detected using a novel spatio frequency algorithm [10]. The coordinates of these calibration fiducials, $\{(x_i, y_i)\}_{i=1}^{17}$ are then extracted from the C-Arm fluoroscopic image. Prior to using EnPrO, a comprehensive analysis of fiducials was conducted to evaluate the individual contribution of the fiducials to the RPE. As shown in Fig. 2, after obtaining the fiducial weights, the weights are sorted in ascending order. As the

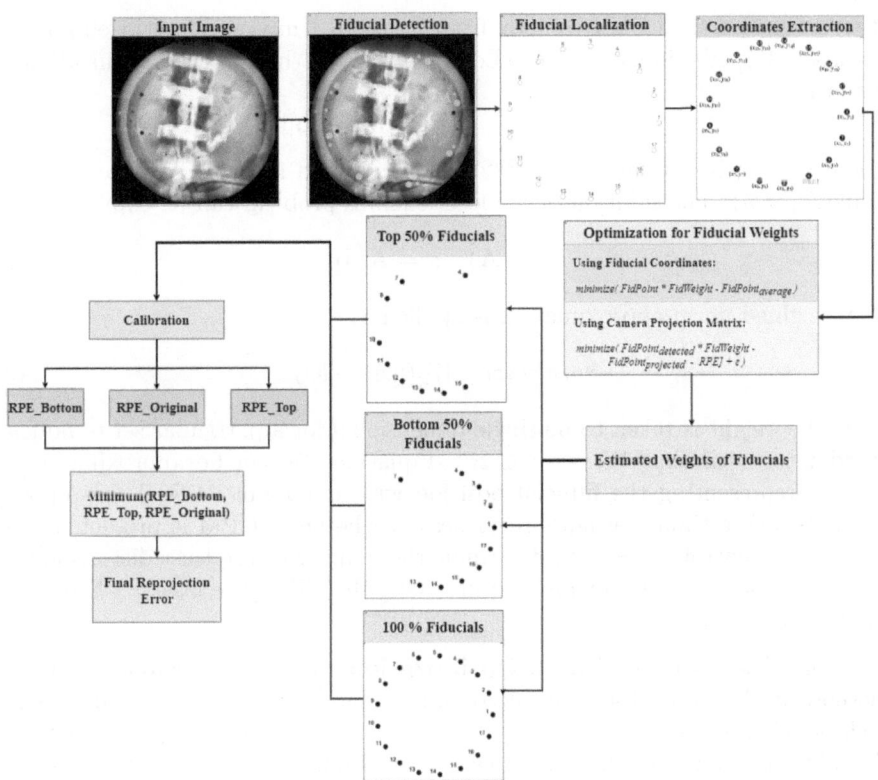

Fig. 2. Schematic representation of the proposed EnPrO: C-Arm input image is computationally processed for fiducial extraction which is then utilized in weighted norm optimization. As only less than half of the detected fiducials are required for camera calibration matrix computation, we split the fiducials into top and bottom fiducials and do the RPE comparison.

minimum number of fiducials required for proper calibration is only 50%, we can now choose to keep the best half of the detected fiducials. Consequently, the ten fiducials with the highest weights and the ten fiducials with the lowest weights are selected. The RPE obtained when all distortion fiducials are used in calibration is compared with the errors obtained when the top ten fiducials and the bottom ten fiducials, along with the significant fiducials, are used for calibration. The fiducials that give the minimum error are the optimal fiducials for that image.

3.1 Optimization for Fiducial Weights

This paper proposes two methods under EnPrO, an optimization methodology trying to tie the fiducial coordinates to the RPE. The intent is to make the best possible choice of the fiducials by the matrix problem abstraction as a set of error choices. The CVX toolbox from Stanford University is utilized for this purpose [4,5]. We implement the Least Squares method in the optimization of the form

$$minimize \quad f(x) = ||\mathbf{A}x - \mathbf{b}||^2 \tag{1}$$

where $\mathbf{x} \in \mathbb{R}^n$ is obtained, $\mathbf{A} \in \mathbb{R}^{m \times n}$ is skinny and full rank (i.e., $m \geq n$ and Rank($\mathbf{A}$) = n). The solution of the least-squares problem can be expressed as:

$$(\mathbf{A}^T\mathbf{A})^{-1}x = \mathbf{A}^T\mathbf{b} \tag{2}$$

The weighted norm approximation is applied as

$$minimize \quad W_x(||Ax - b||) \tag{3}$$

Here, the weight is taken to be single dimensional for a particular set of fiducial coordinates such that $W_x \in x \in \mathbb{R}^n$. Equation (3) can be approximated as $W_x||A||$ representing the fiducial position with minimum RPE. The proposed methods may estimate weights regardless of whether a CPM is present. In the former, the centroid is used to determine the point with the least distortion and RPE, whereas in the latter method involving the CPM, the least RPE point is computed.

Fiducial Co-ordinates Based Optimization. The objective function in this method involves only the fiducial coordinates and the average coordinates for each fiducial. Thus, the weights are determined for the x and y coordinates of the fiducials using the following objective function:

$$\begin{bmatrix} (x_1^1, y_1^1) & \cdots & (x_1^m, y_1^m) \\ \vdots & \ddots & \vdots \\ (x_n^1, y_n^1) & \cdots & (x_n^m, y_n^m) \end{bmatrix} \begin{bmatrix} (w_{x1}, w_{y1}) \\ \vdots \\ (w_{xn}, w_{yn}) \end{bmatrix} - \begin{bmatrix} (x_{1average}, y_{1average}) \\ \vdots \\ (x_{naverage}, y_{naverage}) \end{bmatrix} \tag{4}$$

where,
   $A = [(x_i^1, y_i^1) \cdots (x_i^m, y_i^m)]$, $x = [(w_{xi}, w_{yi})]$ and $b = (x_{iaverage}, y_{iaverage})$ ($i = 1, \ldots, n$)

Here, (x, y) denotes the coordinates of the fiducials obtained during the fiducial extraction algorithm. The average of each fiducial's x and y coordinates are represented by $x_{1average}, \ldots, x_{naverage}$ and $y_{1average}, \ldots, y_{naverage}$ respectively. The minimum RPE fiducial position is considered to be the centre of all the fiducial coordinate predictions taken individually as that point has the least distortion. The weights are now estimated to minimize the difference between actual erroneous fiducial coordinates to fiducial coordinates which have minimum RPE as shown in Algorithm 1.

$$\hat{x}_w = argmin(||A_{ij}x_w - b_{j\,min}||^2) \tag{5}$$

The above weights are estimated to optimize the coordinates together and using these weights, we can infer which fiducials have more error and hence contribute to reduction in accuracy. Since one of the primary reasons for the fiducial errors is due to the C-Arm device-induced distortion, the proposed weights would be apt for a particular C-Arm given a suitable amount of data is collected from that C-Arm. The above approach works on the intuition of using the centroid as the minimum RPE fiducial coordinate.

Camera Projection Matrix (CPM) Based Optimization. Another approach we tried in this paper is to utilize the reprojected fiducial coordinate using CPM as

$$F'(x, y) - F(x, y) = RPE \tag{6}$$

where F(x, y) and F'(x, y) are the original points and reprojected points respectively. This should provide a more robust approximation than before and can be modelled as weighted robust minimization as shown in Algorithm 2,

$$\hat{x}_w = argmin(||A_{ij}x_w - ||P^{proj} - RPE + \varepsilon(P^{proj})||||)^2 \tag{7}$$

where P^{proj} represents the reprojected point vector $\mathbb{R}^n$.

The objective function in this method incorporates the detected points, projected points and the original RPE. Detected points are obtained during the detection of fiducials. The projected points and RPE are obtained during the calculation of a camera projection matrix using Direct Linear Transformation (DLT) [1].

$$\begin{bmatrix} (x_1^1, y_1^1) \cdots (x_1^m, y_1^m) \\ \vdots \ddots \vdots \\ (x_n^1, y_n^1) \cdots (x_n^m, y_n^m) \end{bmatrix} \begin{bmatrix} (w_{x1}, w_{y1}) \\ \vdots \\ (w_{xn}, w_{yn}) \end{bmatrix} - \begin{bmatrix} (x_1', y_1') \\ \vdots \\ (x_n', y_n') \end{bmatrix} - RPE + \begin{bmatrix} \varepsilon_1 \\ \vdots \\ \varepsilon_n \end{bmatrix} \tag{8}$$

where,

A $= [(x_i^1, y_i^1) \cdots (x_i^m, y_i^m)]$, x $= [(w_{xi}, w_{yi})]$ and b $= [(x_i', y_i') - RPE] + [\varepsilon_i]$ $(i = 1, \ldots, n)$

Here, (x, y) is the detected point and (x', y') is the projected point. RPE is the reprojection error previously obtained during calibration. The above minimization becomes more robust as the $RPE_{minimum}$ is computed rather than being picked on using intuition in the previous method.

Algorithm 1. Optimize Fiducial Weights x_n: Using Fiducial Coordinates

---

Require: $A_{m \times n}$ and b_m

Use Weighted Norm: $W_x(||Ax - b||)$

$||W_x A|| ||W_x x|| - ||W_x b||$

$||W_x A|| = A_{ij} =$ Fiducial Position

$||W_x b|| = b_{j\,min} =$ Fiducial Position with Minimum RPE

$||W_x x|| = x_w =$ Fiducial Weight

if $A_{ij}x_w - b_{j\,min} = 0$ **then**

$\quad x_w = $ [Unity Matrix]

else

$\quad \hat{x}_w = argmin(||A_{ij}x_w - b_{j\,min}||^2)$

$\quad$ Converge at $A_{ij}^T A_{ij} x_w = A_{ij}^T b_{j\,min}$

end if

Algorithm 2. Optimize Fiducial Weights x_n: Using Camera Pojection Matrix

---

Require: $A_{m \times n}$ and b_m

   Use Weighted Norm: $W_x(|||Ax - b||)$

   $||W_x A||||W_x x|| - ||W_x b||$

   $||W_x A|| = A_{ij} = $ Fiducial Position

   $||W_x b|| = b_{j\,min} = $ Fiducial Position with Minimum RPE

   $||W_x x|| = x_w = $ Fiducial Weight

   $b_{j\,min} = ||P^{proj} - RPE + \varepsilon(P^{proj})||$

   $P^{proj} = $ Projected Points

   RPE = Reprojection Error

   $\varepsilon = $ Robust Approximation Term

   if $A_{ij} x_w - b_{j\,min} = 0$ **then**

      $x_w = [$Unity Matrix$]$

   else

      $\hat{x}_w = argmin(||A_{ij} x_w - b_{j\,min}||^2)$

        $= argmin(||A_{ij} x_w - ||P^{proj} - RPE + \varepsilon(P^{proj})|||||)^2$

      Converge at $x_w = (A_{ij})(P^{proj} - RPE + \varepsilon(P^{proj}))$

   end if

In Equations (4) and (8), n and m represent the number of fiducials and the number of images respectively. The weights $w_{x1}, w_{x2}, \ldots, w_{xn}$ and $w_{y1}, w_{y2}, \ldots, w_{y_n}$ represent the multiplication factors that the optimization algorithm gives as output for the x and y coordinates respectively, to keep the error minimum. The weight of the fiducial is the average of these weights obtained for the x and y coordinates separately. The weights for the individual coordinates are obtained by solving the equations for all the fiducials detected in every image.

4 Results and Discussion

Dataset: The proposed EnPrO was modelled on 309 C-Arm fluoroscopic images captured from two different C-Arms. The IGSS in-house design had 17 fiducials reserved for calibration and 64 fiducials reserved for distortion. The weights were computed using only the calibration fiducials whereas the RPE metric was evaluated over the entire distortion and calibration fiducials to make it more generalized across the image. The absence of a particular fiducial due to non-detection or occlusion usually leads to an increase or decrease in RPE. Since for IGSS, the medical-grade accuracy is less than 2 mm, it becomes important to analyse these variations in RPE using the proposed EnPrO.

Study of Fiducials: From the RPE analysis performed on all images, the 4th, 5th and 7th fiducials were found to be significant, as shown in Fig. 3(a). The study was further narrowed to images in which all the calibration fiducials were detected (74 images). As described in Fig. 3(b), images with all the calibration fiducials showed that removing the 4th and 7th fiducials significantly increased the RPE, thus identifying them as the most significant fiducials. In contrast, removing the 6th and 9th fiducials minimized the error, classifying them as non-significant. The images used in this analysis were further classified based on the C-Arm used for acquisition. When C-Arm 1 was used for image

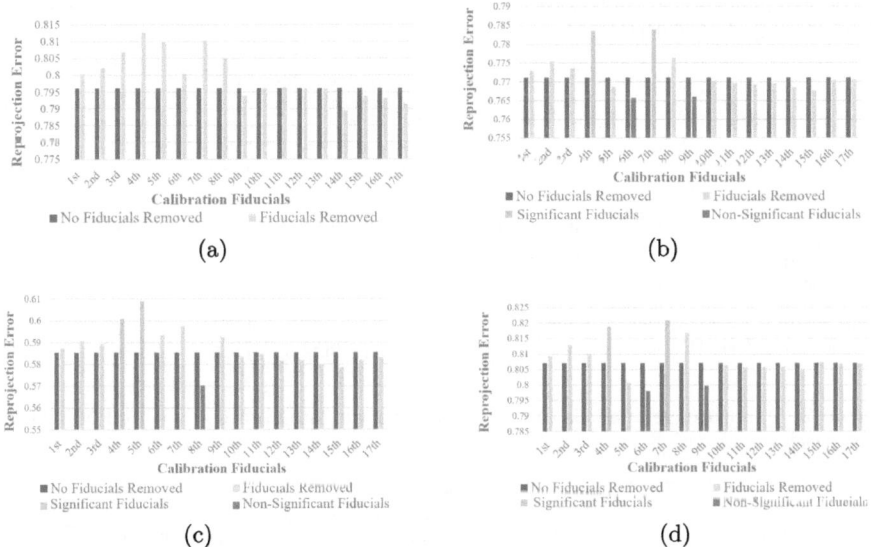

Fig. 3. RPE Analysis of the Calibration Fiducials: (a) All dataset images, (b) Images with full Calibration Fiducials, (c) Images with full Calibration Fiducials acquired using C-Arm 1, (d) Images with full Calibration Fiducials acquired using C-Arm 2

acquisition (13 images), the 4th and 5th fiducials were significant. The 8th fiducial was non-significant, as shown in Fig. 3(c). In images acquired using C-Arm 2 (61 images), depicted in Fig. 3(d), the 4th and 7th fiducials were significant, whereas the 6th and 9th fiducials were non-significant. From this study, it was predicted that including the 4th and 7th fiducials and excluding the 6th and 9th fiducials during calibration would reduce the RPE.

Results Using EnPrO: Using EnPrO, we obtained the fiducial coordinates weights using the above-mentioned fiducial coordinates method and the CPM on images from two different C-Arms manufacturers. Ten fiducials with the highest and lowest weights were chosen for calibration and the RPE obtained in this case was also compared with the RPE obtained when the significant fiducials were included and the non-significant fiducials were excluded in the same cases. In the proposed EnPrO for all images, the minimum RPE from the above scenarios was chosen as the final RPE. Table 1 represents the fiducial weights obtained using fiducial coordinates in EnPrO where the RPE decrease of 6.96 % is achieved. When the fiducial weights were obtained using the CPM, the RPE decreased by 8.36% and the results are presented in Table 2. The ε varied from -200 to +200 in steps of 10.

Results from Different C-Arms: As demonstrated in Fig. 4, the RPE measured with the proposed EnPrO is compared on images acquired using C-Arm 1 and C-Arm 2 taken separately and together, the fiducials chosen using the weights obtained from the fiducial coordinates minimize the RPE 4(a) whereas in images acquired using C-Arm 2, using the CPM gives weights that help choose the optimal fiducials 4(b).

Table 1. Quantitative analysis of the RPE using weights obtained with fiducial coordinates method of EnPrO

Experiments		Initial RPE	Fiducials chosen for calibration based on their weights				Final RPE	Decrease in RPE (%)
			Bottom 10	Bottom 10 Include 4, 7 Exclude 6, 9	Top 10	Top 10 Include 4, 7 Exclude 6, 9		
All images with 17 Fiducials	Average	0.7742	0.7829	0.7632	0.8947	0.7835	0.7203	6.96
	SD	0.1018	0.143	0.1169	0.1337	0.1167	0.0986	
Images with 17 fiducials acquired using C-Arm1	Average	0.5855	0.5458	0.6112	0.8625	0.6512	0.5579	4.72
	SD	0.0546	0.0661	0.0975	0.1774	0.103	0.0381	
Images with 17 fiducials acquired using C-Arm2	Average	0.8071	0.8243	0.7897	0.9004	0.8066	0.7487	7.24
	SD	0.066	0.1083	0.0982	0.1236	0.1028	0.0758	

Table 2. Quantitative analysis of the RPE using weights obtained with CPM method of EnPrO

Experiments		Initial RPE	Fiducials chosen for calibration based on their weights				Final RPE	Decrease in RPE (%)
			Bottom 10	Bottom 10 Include 4, 7 Exclude 6, 9	Top10	Top10 Include 4, 7, Exclude 6, 9		
All images with 17 Fiducials	Average	0.7742	0.7520	0.7256	0.8296	0.7868	0.7094	8.36
	SD	0.1018	0.1175	0.0971	0.1511	0.1118	0.0966	
Images with 17 fiducials acquired using C-Arm1	Average	0.5855	0.6217	0.5839	0.6859	0.6353	0.5472	6.54
	SD	0.0546	0.1345	0.0714	0.0927	0.0840	0.0423	
Images with 17 fiducials acquired using C-Arm2	Average	0.8071	0.7748	0.7503	0.8547	0.8133	0.7378	8.59
	SD	0.0660	0.0979	0.0779	0.1452	0.0935	0.0725	

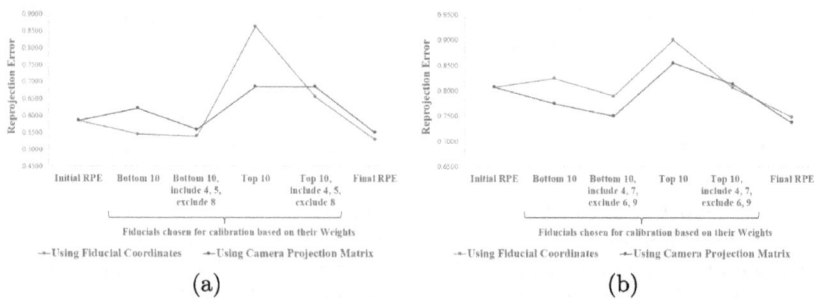

(a)   (b)

Fig. 4. Performance Comparison of the EnPrO for the RPE metric on (a) images with 17 Fiducials acquired using C-Arm 1, (b) images with 17 Fiducials acquired using C-Arm 2

Phantom Study: To validate the EnPrO method, a phantom clinical research was conducted. The fiducials were chosen based on the optimization weights and tracking results provided in shown in Fig. 5. As illustrated, IGSS tracking for a pedicle screw point in the L4, L5 spine is accurate, as indicated by the colour code, with the yellow colour of the tracking tool representing accuracy between 1 mm and 2 mm and the green colour representing accuracy less than 1 mm.

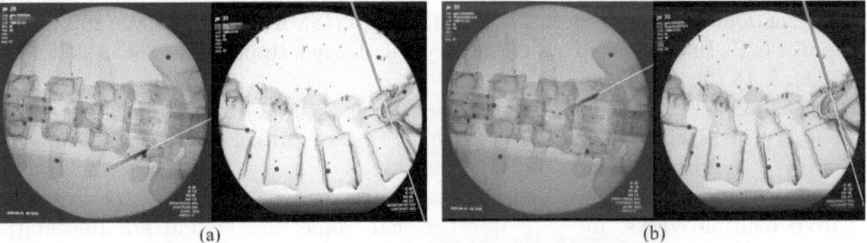

(a)   (b)

Fig. 5. Results using proposed EnPrO on clinical phantom study. The results show the anatomy tracking on (a) AP and (b) LP images based on weighted fiducial selection

Both the above approaches allow us to understand the contribution of each fiducial in IGSS. Using the computed weights, we can improve the precision by avoiding the fiducials contributing to more error. The weights can also be generalized across the device C-Arms though we find the weights at this stage to be non-device agnostic.

5   Conclusion

Prior to the proposed EnPrO, a fiducial-based study for RPE comparison was conducted which proved that each fiducial contributed differently to the RPE. The presence of significant fiducials during calibration minimizes the error and thus ensures higher accuracy by better mapping between the 3D patient anatomy and the 2D image for IGSS. Optimal fiducials that minimise the RPE can be chosen based on their weights obtained using EnPrO. Fiducials were chosen based on weights obtained using the fiducial coordinates and the CPM clearly showed an RPE decrease of 6.96% and 8.36% respectively. Future studies would include conducting comprehensive clinical research using phantom and cadaver trials, as well as linking the RPE effect to clinical tracking accuracy.

References

1. Bertelsen, A., Garin-Muga, A., Echeverría, M., Gómez, E., Borro, D.: Distortion correction and calibration of intra-operative spine X-ray images using a constrained DLT algorithm. Comput. Med. Imaging Graph. **38**(7), 558–568 (2014)
2. El hazzat, S., Saaidi, A., Karam, A., Satori, K.: Incremental multi-view 3D reconstruction starting from two images taken by a stereo pair of cameras. 3D Res. **6**(1–18) (2015)
3. Foley, K.T., Simon, D.A., Rampersaud, Y.R.: Virtual fluoroscopy: computer-assisted fluoroscopic navigation. Spine **26**(4), 347–351 (2001)
4. Grant, M., Boyd, S.: Graph implementations for nonsmooth convex programs. In: Blondel, V., Boyd, S., Kimura, H. (eds.) Recent Advances in Learning and Control, pp. 95–110. Springer-Verlag Limited, Lecture Notes in Control and Information Sciences (2008)

5. Grant, M., Boyd, S.: CVX: Matlab software for disciplined convex programming, version 2.1. https://cvxr.com/cvx (Mar 2014)
6. Heemeyer, F., Choudhary, A., Desai, J.P.: Pose-aware C-arm calibration and image distortion correction for guidewire tracking and image reconstruction. In: 2020 International Symposium on Medical Robotics (ISMR), pp. 181–187 (2020)
7. Holly, L.T.: Image-guided spinal surgery. Int. J. Med. Robot. Comput. Assist. Surg. $2(1)$, 7–15 (2006)
8. Hsieh, J.C., et al.: Accuracy of intraoperative computed tomography image-guided surgery in placing pedicle and pelvic screws for primary versus revision spine surgery. Neurosurg. Focus $36(3)$, E2 (2014)
9. Mai, F., Hung, Y.: Augmented lagrangian approach for projective reconstruction from multiple views. In: 18th International Conference on Pattern Recognition (ICPR'06), IEEE (2006)
10. Purayath, A., Maik, V., Abhilash, Lakshmanan, M., Sivaprakasam, M.: A novel spatio^2-frequency blob detection algorithm for enhancing precision in image guided surgery (2024)
11. Sommer, F., Goldberg, J.L., McGrath, L., Kirnaz, S., Medary, B., Härtl, R.: Image guidance in spinal surgery: a critical appraisal and future directions. Int. J. Spine Surg. $15(s2)$, S74–S86 (2021)
12. Tungadio, D.H., Numbi, B.P., Siti, M.W., Jordaan, J.A.: Weighted least squares and iteratively reweighted least squares comparison using particle swarm optimization algorithm in solving power system state estimation. In: Africon, IEEE (2013)
13. Zhang, X., Zheng, G.: Robust automatic detection and removal of fiducial projections in fluoroscopy images: an integrated solution. In: 30th Annual International Conference of the IEEE Engineering in Medicine and Biology Society, IEEE (2008)
14. Zhang, Z.: A flexible new technique for camera calibration. IEEE Trans. Pattern Anal. Mach. Intell. $22(11)$, 1330–1334 (2000)

Patient-Specific 3D Burn Size Estimation

Kim-Ngan Nguyen[1]([✉]), Han Ching Yong[2], and Terence Sim[1]

[1] National University of Singapore, Singapore, Singapore
knnguyen@nus.edu.sg, tsim@comp.nus.edu.sg
[2] Defence Science and Technology Agency, Singapore, Singapore
hanchingyong@gmail.com

Abstract. In the treatment of burn wounds, an accurate estimation of the ratio of the wound's area to the total body surface area (%TBSA) is crucial. Medical doctors use %TBSA as one of the factors to determine the initial treatment, track the healing process, and devise subsequent treatments. Existing works estimate the %TBSA using generic human models with predefined percentages for each body part. These generic human models do not consider the patients' actual body shape. Furthermore, the estimation of wound size depends greatly on the doctor's experience, resulting in inaccurate estimated %TBSA. This work addresses the problem by using 3D modeling techniques and machine learning to give a better estimation of %TBSA for each individual patient. We do this by training a regressor to predict the patient's body surface area from images of his face and hand, and by using a portable 3D scanner to determine the wound's area. Our method estimates %TBSA with an average error of 8.5%, which is a huge improvement over the 140% error produced by existing methods. Our method is also easily accessible since it uses commercial-off-the-shelf (COTS) devices. This makes it practical for anyone, even those without medical knowledge, to use. Project page: https://github.com/nganntk/BurnWounds.

Keywords: Burn wounds · Wound size estimation · Patient-specific

1 Introduction

The severity of the burn wounds is assessed based on the cause of the wound (chemical, thermal, etc.), and the characteristics of the wound such as the depth of the wound and the ratio of the wound surface area (WSA) and total body surface area (TBSA), i.e., $\%TBSA = \text{WSA}/\text{TBSA}$. Among these characteristics, the %TBSA is the most difficult to assess by paramedics in the initial assessment, i.e., outside of burn centers, because the wounds may have irregular shapes and spread over a large area of the patient's body. Over- and under-estimation of the %TBSA will cause over- and under-resuscitation of burn fluid, which may lead to serious complications such as edema formation, acute kidney injury, etc.

This work was done when Han Ching Yong was in NUS.

© The Author(s), under exclusive license to Springer Nature Switzerland AG 2024
K. Drechsler et al. (Eds.): CLIP 2024, LNCS 15196, pp. 53–62, 2024.
https://doi.org/10.1007/978-3-031-73083-2_6

Thus, proper initial assessment of %TBSA by the paramedics is crucial to avoid such undesired outcomes and to determine the transferring of a burn patient to a specific burn center [15].

To estimate the %TBSA, existing works adopt either generic-%TBSA or specific-%TBSA. Generic-%TBSA uses templates with predefined percentages of each body part to estimate %TBSA. These templates were built based on either clinically-defined percentage charts or 2D/3D models of the human body. Chart-based and 2D-model methods, such as the Lund and Browder Chart, are commonly used in practice due to their simplicity, but they do not account for the curvature of the human body [2,11,19,23]. On the other hand, 3D-model methods capture the curvature of the human body, however, they need either manual marking of the wound on the 3D model [14] or utilize a Kinect scanner which is error-prone and impractical [18].

Specific-%TBSA methods estimate WSA and TBSA separately to compute %TBSA. They measure WSA by either placing a measurement tool, i.e., grid paper or ruler, onto the wound, or using a camera to capture the entire wound in a single image [17,22]. Then, they estimate TBSA using the formula TBSA $=$ Height^a × Weight^b × c , where a, b, c are constants that adjust the importance of height and weight [3,7–9,13]. These constants were derived from empirical data, and do not consider biometric features that can affect the body shape such as the age and gender of the patient.

The main shortcomings of existing methods are their impracticality, slowness, and the use of generic models that do not consider the differences in the curvature of different body parts. For example, burn size estimation requires burn wound segmentation as a prerequisite step, even though 2D images do not capture the accurate 3D shapes of burn wounds, burn wound segmentation models are still developed using 2D images as input. In this paper, we propose a novel method that overcomes the above shortcomings. Our main contributions are:

- A new dataset of accurate 3D body scans of 21 healthy human subjects, along with their face and hand images. This will be publicly released to the research community.
- A novel method to estimate total body surface area that is patient-specific rather than generic.
- A novel method to estimate wound surface area using a COTS scanner that is patient-specific, and that accounts for body curvature.

2   Methods

Our work falls under the specific-%TBSA approach, we propose a novel and practical method for estimating the WSA (Sect. 2.1) and TBSA (Sect. 2.2) along with a method for automatically measure %TBSA with accessible devices.

2.1   Estimating WSA

To estimate WSA, we use a COTS scanner, namely, an iPhone, that outputs a 3D colored mesh P of the scanned burn wound region. Vertex v_i of P includes the

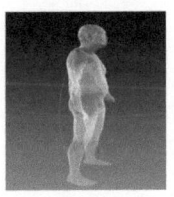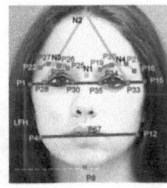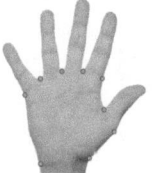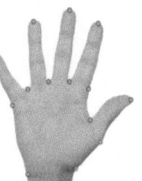

Fig. 1. Left to right ((a) - (e)): sample of 3D body scan, face shape, face geometry [24], palm only features, and palm with fingers features.

position and the color of v_i. The WSA is computed using the following procedure (Fig. 2a):

1. Remove duplicate, non-manifold vertices and faces on mesh P.
2. Compute the mesh parameterisation U of P using Least Squares Conformal Maps [12] that maps each vertex v_i in the mesh P to a 2D point $U_i = (u, v)$.
3. Render U to a 2D image I of a fixed size, i.e., 512×512.
4. Perform burn wound segmentation on I to obtain a segmentation mask of the burn wound I'.
5. Map the segmented burn wound region in I' to P to obtain the wound mesh P'. Specifically, for each pixel in I', find its approximated vertex v_i in P. Then, take all the neighboring vertices of v_i that are within a certain radius r, i.e., $r = 0.02$ mm.
6. Compute WSA $= \sum_t A_t$ for all faces in P', where A_t denote the area of a face at index t in P'.

For the segmentation in step 4, we used a recent burn wound segmentation method called DeepASPP [4]. Note that this may be replaced by any other segmentation method.

2.2 Estimating TBSA

Inspired by existing works that estimate a human body mass index (BMI) from human face photo [6,24], our proposed method uses face and palm features extracted from their respective photographs to build a linear regressor that estimates the TBSA of a person. To achieve this, the proposed method is composed of three stages, namely (1) data normalization, (2) feature selection, and (3) training a linear regressor. Note that we also developed TBSA estimation models using other regression models, but linear regression gives the best performance.

First, shape normalization using stable points is applied to ensure uniform size for all face and palm photographs. Stable points are the three points with a constant distance ratio regardless of the subject's expression or skin movement. For the face, they are the center of the eyes and the nose. For the palm, they are the bottom corners of the index and pinky fingers and the bottom left corner of the palm. A reference image I_r is manually chosen, and all other images are aligned using affine transformation to match the stable points in I_r, ensuring similar sizes of faces and palms across images.

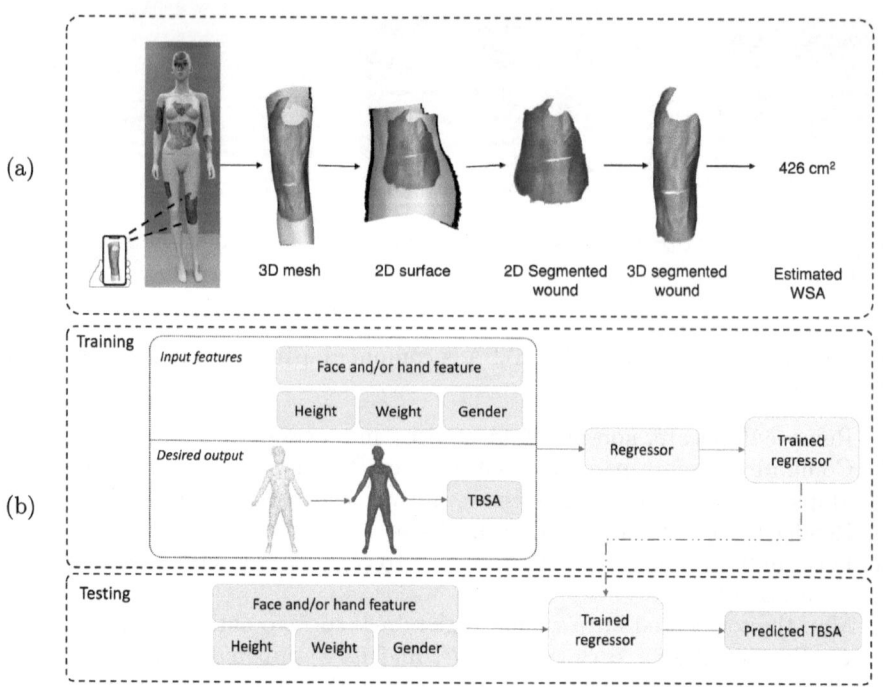

Fig. 2. Proposed method for burn size estimation. (a) Obtain WSA using an iPhone. (b) Regression model to estimate TBSA from biometrics features.

Second, several features are selected for both the face and palm. For the face, two types of facial features are computed from facial landmarks, namely face-shape and face-geometry. The face-shape features include the outline of the face, the nose, and the mouth which should reflect the person's body shape (Fig. 1b). The face-geometry features include the seven physiology features that have been proven to be able to give clues on a person's weight and BMI [5,16,24]. Specifically, the face-geometry features are: (1) Cheekbone width over jaw width, (2) Cheekbone width over upper facial height, (3) Perimeter over area of the face, (4) Average size of both eyes, (5) Lower face height over entire face height, (6) Face width over lower face height, (7) Average distance between eyebrows and the upper edge of eyes. For the palm, two types of features are used: (1) palm without fingers (Fig. 1d) and (2) palm with fingers (Fig. 1e). (1) includes the creases around the palm, whereas (2) adds in the tip of the fingers. Next, the visual features, height, and weight are normalized. The normalization is necessary to ensure the convergence of the linear regressor during training.

Finally, the last stage is to train a linear regressor on the selected features. Given a dataset of n subjects, each subject $i(i = \overline{1, n})$ has a set of features F that includes a face feature f_i, a palm feature p_i, height h_i, weight w_i, and a TBSA t_i. A linear regressor is trained that takes one or more features in F and predicts the estimated TBSA of the subject t'_i (Fig. 2b). Since the body compositions of males and females are usually quite different, we use two separate regressors, one for each gender. During inference, each regressor will then take a combination of visual features, i.e., face and/or palm features, height, and weight to form the feature vector. To evaluate the accuracy of the proposed model, we use leave-one-out cross validation due to a limited number of data samples.

3   Experiments

3.1   Data Collection

We collected and used two datasets: (1) 3D-Healthy-Body-Scan for developing the TBSA prediction model and (2) Burn-Wound-Scan for evaluating the proposed method for burn size estimation.

For the 3D-Healthy-Body-Scan dataset, we collected 3D body scans of 21 *healthy* human subjects (10 females, 11 males, 21 - 50 years old) and their biometrics features, i.e., face photograph, palm photograph, weight, height, gender, and age group. This data collection is a one-time effort used to train a regressor. In practice, the 3D body scan of burn patients is not required. To train the regressor, the TBSA of each participant is needed. TG3D Studio's Scanatic 360 Body Scanner [20] is used to obtain the body shape as a point cloud (Fig. 1a). Then, Poisson surface reconstruction [10] is used to generate a 3D mesh. Finally, the TBSA is computed as the sum of areas of all faces in the 3D mesh.

For the Burn-Wound-Scan dataset, due to the lack of publicly available burn wound images especially in 3D and COVID restrictions, we devised a prolific way to replicate the burn wound on patients to evaluate the performance of the proposed method. To achieve that, we created artificial wounds by using two mannequins (1 male, 1 female) and 13 wound images from the dataset in [4]. We note that this practice of using mannequins with artificial burn wounds is commonly used in burn wound analysis literature as well as for medical teaching and research purposes. The wounds were placed on various body parts with different levels of curvature such as the torso which has low curvature and the arm which has high curvature. To scan the wound, we considered two scanners, Artec Eva [1] and iPhone 12's True Depth camera. Artec Eva uses structured light while the iPhone's TrueDepth camera uses an infrared sensor to create a point cloud with color information. For the iPhone, we used Heges API to retrieve the most accurate point cloud from the iPhone's sensor [21]. The scanning time for a burn wound is <5 min for the Artec and <1 min for the iPhone.

3.2   Performance Metric and Evaluation Setup

Percentage error is used to evaluate the accuracy of the proposed method in computing WSA, TBSA, and %TBSA. Given the estimated value v_E and the ground-truth value v_A, the percentage error δ is computed as follows:

$$\delta = \frac{|v_E - v_A|}{v_A} \times 100 \tag{1}$$

The ground-truth TBSA of the two mannequins were obtained in a similar way to those of participants (Sect. 3.1), but the Artec scanner was used instead. The acquisition of ground-truth WSA was more involved, which will be described in the next paragraph. The estimated TBSA and WSA are computed as described in Sect. 2.2 and Sect. 2.1, respectively. Finally, the estimated and ground-truth %TBSA is computed from TBSA and WSA.

Due to the wounds' irregular shapes, it is not trivial to get the actual wound area. Therefore, we obtain the wound area by taking a picture of each wound with a reference object of a known size (i.e., a green sticker with a diameter of 12 mm). Then, the camera is calibrated using the standard calibration method [25] to remove any effects from lens distortion. Finally, the area of the wound in meter W_m is computed from the area of the wound in pixel W_p, the area of the reference object in pixel R_p and meter R_m using Eq. 2. This area is the pseudo ground truth of the wound and is considered the ground truth wound size for the following experiments.

$$W_m = \frac{W_p * R_m}{R_p} \tag{2}$$

The performance evaluation of the proposed method was done by comparing its metrics with three existing works, namely (1) Rule of 9 s, (2) Lund and Browder chart, and (3) Mersey Burns. To obtain the results of the previous works, two researchers separately computed the %TBSA using the previous work the average of the computed %TBSA between the two researchers is used for comparison. We also manually segmented the burn wound to obtain the upper-bound performance of our method.

4   Results and Discussions

4.1   Percentage Error of the TBSA

Table 1 summarizes the percentage error of the TBSA for each gender group with different combinations of visual features. The average percentage error is less than 10 % for both genders. The model always performs slightly better in the male group compared to the female group, except when only the face feature is used. The smallest average errors for female and male groups are 3.46% and 4.00%, respectively, and they are obtained with face shape as the visual feature for females and palm region as the visual feature for males.

Table 2 compares the percentage error of the proposed model with those of the previous works. The table shows that the proposed model with face shape feature, i.e., lowest error, gives comparable error for females and males to those of the previous works. The estimated TBSA of the male and female mannequins are 1.57 m^2 and 1.83 m^2, respectively. We note that previous works have also used mannequins to aid the development of burn size estimation methods.

Table 1. Percentage error (%) of TBSA with different visual features.

Models	Female	Male
Face shape (FS)	**3.46**	5.06
Face geometry (FG)	3.73	5.82
Palm only	6.28	**4.00**
Palm + fingertips	6.79	4.06
FS + palm	7.70	6.00
FS + palm + fingertips	9.14	5.91
FG + palm	6.18	4.42
FG + palm + fingertips	3.73	5.82

Table 2. Comparison of percentage error (%) of TBSA

Methods	Female	Male
Bois	1.87	2.90
Boyd	63.15	63.62
Haycock and Schwarz	3.02	3.53
Gehan and George	3.03	3.69
Mosteller	2.57	3.23
Proposed	3.46	5.06

4.2 Percentage Error of the WSA

The performance of the WSA is highly dependent on the wound segmentation model. Current methods in burn wound segmentation focus on 2D burn wound segmentation, however, to get an accurate estimation of the wound size, we need to use 3D scans to account for different curvatures of the wound surface. Furthermore, 2D burn wound segmentation models are often trained on images taken in good lighting conditions and are not tested on a variety of lighting conditions in real life. Therefore, in this work, we evaluated our method with the wounds either manually segmented or by using DeepASPP. The method's performance when the wound is manually segmented is the upper-bound performance, while the model performance when using DeepASPP served as a baseline of the current state-of-the-art method for 2D burn wound segmentation when applied to real-life scenarios. The results showed that when the wounds were manually segmented, both Artec and Heges produced a small average percentage error, i.e., < 10%. However, when the wounds were segmented using DeepASPP, the percentage error of both Artec and Heges increased four times. Artec's average percentage error increases from 6.00% to 25.02%, while the average error for Heges increases from 7.62% to 31.62%. DeepASPP was the SOTA burn wound segmentation model, however, it cannot perform well under real life scenario with varying lighting conditions. Thus, our work aims to bring focus on the gap between the performance of burn wound segmentation model performance on 2D images compared to its performance in practical conditions.

4.3 Percentage Error of the %TBSA

Table 3 compares the percentage error of %TBSA between the proposed model and the previous works, i.e., chart-based and 2D model-based methods. By accounting for the curvature of the specific patient, especially when computing the WSA, the proposed method achieves a significantly lower percentage error of %TBSA compared to previous works. The average errors of the proposed method

Table 3. Comparison of percentage error of %TBSA between the previous works (3 left columns) and the proposed method (4 right columns). (F) means female, (M) means male. The previous works include: (Ro9) Rule of 9 s, (LB) Lund and Brow- der chart, and (MB) Mersey Burns. For the proposed method, different variant of the proposed methods are compared: (ArtecM) Artec-manual, (ArtecD) Artec-DeepASPP, (HegesM) Heges-manual, (HegesD) Heges-DeepASPP. All values are in percentage.

Wound position	Ro9	LB	MB	ArtecM	ArtecD	HegesM	HegesD
(F) arm large	20.26	36.66	33.02	6.09	4.34	3.24	14.61
(F) arm small	51.22	46.34	17.07	8.00	8.00	1.27	35.71
(F) foot	26.88	384.66	206.75	22.09	65.03	20.25	71.78
(F) hand	98.10	161.72	134.38	25.78	0.00	21.09	66.41
(F) headback	293.36	205.94	188.46	0.52	12.59	9.79	116.08
(F) headfront large	199.54	132.92	150.57	19.82	30.98	9.11	0.91
(F) headfront small	241.73	166.19	223.74	0.72	46.04	4.32	50.36
(F) leg large	3.20	0.61	19.33	11.91	37.85	12.36	37.56
(F) leg small	18.80	13.58	82.77	2.35	29.63	0.26	13.06
(F) torsoback large	229.52	122.07	77.66	0.82	4.61	2.80	30.34
(F) torsoback small	193.05	112.01	167.73	3.62	12.30	0.43	14.47
(F) torsofront large	276.28	171.76	52.03	0.53	19.50	6.69	16.42
(F) torsofront small	217.04	129.19	190.62	1.72	16.91	1.32	21.27
(M) arm large	21.92	20.76	34.70	4.78	2.18	7.71	28.57
(M) arm small	88.44	88.44	69.60	9.42	15.96	12.44	50.50
(M) foot	56.00	276.00	260.00	10.40	172.00	16.00	78.40
(M) hand	155.10	185.71	308.16	7.14	47.96	15.31	19.39
(M) headback	452.46	329.69	290.63	6.92	6.25	12.05	2.90
(M) headfront large	286.03	200.74	208.82	6.77	3.24	12.65	1.77
(M) headfront small	287.17	200.89	209.74	13.27	94.69	12.39	2.66
(M) leg large	26.64	29.59	28.75	6.67	16.25	9.46	76.40
(M) leg small	33.90	61.02	94.92	8.31	13.22	10.00	30.17
(M) torsoback large	117.39	56.94	113.25	4.35	1.08	6.13	29.73
(M) torsoback small	226.49	135.08	226.49	7.09	2.99	6.90	0.75
(M) torsofront large	183.42	104.54	67.69	7.46	8.60	9.54	8.17
(M) torsofront small	131.96	67.53	217.87	10.65	16.50	11.00	12.20
AVERAGE	151.38	132.33	141.34	**7.97**	**26.49**	**9.02**	**31.95**

are around 8–9% for manual segmentation and 26–31% for segmentation using DeepASPP. These errors are significantly smaller than those of the previous works, where the average errors are around 130% to 150%. When comparing the average errors between the Artec and Heges, their difference is not significant, i.e., around 1% with manual segmentation. Most importantly, the errors of both

Artec and Heges are less than 10% with manual segmentation, which is about 15 times smaller than those of the previous works.

This proof-of-concept work currently has some limitations. First, we used mannequins because COVID-19 prevented access to real patients. Second, we used fake wounds, resulting in a clear boundary between the wound and the skin which makes it easier to segment compared to real wound. However, this could be addressed by incorporating a more accurate and robust burn wound segmentation model than the SOTA model, i.e., DeepASPP [4]. Finally, we only have a limited dataset of body scans for estimating TBSA. In future steps, we aim to broaden the 3D body scan dataset to encompass diverse age groups and ethnicities. Additionally, we intend to integrate our model into a mobile app and conduct clinical trials with real patients to enhance and validate its performance.

5   Conclusion

This work presents an automated, accurate, and easy-to-use method to estimate the %TBSA with COTS devices. More precisely, we introduce a new dataset that includes 3D body scans and biometric features of 21 healthy human subjects. We also proposed a method that estimates the TBSA from biometrics features and estimates WSA using 3D modeling techniques. We emphasize that each patient only needs to provide (a) a video of the burn wound, (b) a single photo of the face and/or palm, and (c) personal information that consists of height, weight, and gender. No 3D body scan of the patient is needed. This makes our method practical because these inputs are easily acquired from the patient, who may be in great pain because of the burns. Our experiments with different wound sizes, locations, and 3D scanners demonstrate the effectiveness and accuracy of our proposed method as compared to the previous works. These promising outcomes motivate further clinical trials and eventual practical deployment.

Disclosure of Interests. The authors have no competing interests to declare that are relevant to the content of this article.

References

1. Artec3D: Artec eva. https://www.artec3d.com/portable-3d-scanners/artec-eva
2. Barnes, J., et al.: The Mersey burns app: evolving a model of validation. Emerg. Med. J. **32**(8), 637–641 (2015)
3. Boyd, E.: The growth of the surface area of the human body. J. Roy. Stat. Soc. **100**(1), 111 (1937)
4. Chauhan, J., Goyal, P.: Convolution neural network for effective burn region segmentation of color images. Burns **47**(4), 854–862 (2021)
5. Coetzee, V., Chen, J., Perrett, D., Stephen, I.: Deciphering faces: quantifiable visual cues to weight. Perception **39**(1), 51–61 (2010)
6. Dantcheva, A., Bremond, F., Bilinski, P.: Show me your face and i will tell you your height, weight and body mass index. In: ICPR, pp. 3555–3560 (2018)

7. Du Bois, D., Du Bois, E.: A formula to estimate the approximate surface area if height and weight be known. 1916. Nutrition (Burbank, Los Angeles County, Calif.) **5**(5), 303–311; discussion 312–313 (1989)
8. Gehan, E., George, S.: Estimation of human body surface area from height and weight. Cancer Chemother. Rep. **54**(4), 225–235 (1970)
9. Haycock, G., Schwartz, G., Wisotsky, D.: Geometric method for measuring body surface area: a height-weight formula validated in infants, children, and adults. J. Pediatr. **93**(1), 62–66 (1978)
10. Kazhdan, M., Hoppe, H.: Screened poisson surface reconstruction. ACM Trans. Graph. (TOG) **32**(3), 29 (2013)
11. Lund, C., Browder, N.: The estimation of areas of burns. Surg. Gynecol. Obstet. **79**, 352–358 (1944)
12. Lévy, B., Petitjean, S., Ray, N., Maillot, J.: Least Squares Conformal Maps for Automatic Texture Atlas Generation. ACM Trans. Graphics **21**(3), 10 (2002)
13. Mosteller, R.: Simplified calculation of body-surface area. N. Engl. J. Med. **317**(17), 1098 (1987)
14. Parvizi, D., et al.: Burncase 3D software validation study: burn size measurement accuracy and inter-rater reliability. Burns **42**(2), 329–335 (2016)
15. Pham, C., Collier, Z., Gillenwater, J.: Changing the way we think about burn size estimation. J. Burn Care Res. **40**(1), 1–11 (2019)
16. Pham, D., Do, J.H., Ku, B., Lee, H., Kim, H., Kim, J.: Body Mass Index and Facial Cues in Sasang Typology for Young and Elderly Persons. Evid. Based Complement. Altern. Med. **2011**, e749209 (2011)
17. Rueden, C., et al.: Image J2: ImageJ for the next generation of scientific image data. BMC Bioinform. **18**(1), 529 (2017)
18. Sheng, W.B., Zeng, D., Wan, Y., Yao, L., Tang, H.T., Xia, Z.F.: BurnCalc assessment study of computer-aided individual three-dimensional burn area calculation. J. Transl. Med. **12**(1), 242 (2014)
19. Stiles, K.: Emergency management of burns: part 2. https://journals.rcni.com/emergency-nurse/cpd/emergency-management-of-burns-part-2-en.2018.e1815/abs
20. TG3DS: Digital Fashion Evolution with 3D Fashion Technologies
21. Verdaasdonk, R., Liberton, N.: The Iphone X as 3D scanner for quantitative photography of faces for diagnosis and treatment follow-up. In: Optics and Biophotonics in Low-Resource Settings V, vol. 10869, pp. 1086902. SPIE (Mar 2019)
22. Visitrak: Smith and Nephew Healthcare Ltd. VISITRAK Digital. https://www.smith-nephew.com/global/assets/pdf/products/surgical/2-visitrakdigitaluserguide.pdf
23. Wallace, A.: The exposure treatment of burns. Lancet (London, England) **1**(6653), 501–504 (1951)
24. Wen, L., Guo, G.: A computational approach to body mass index prediction from face images. Image Vis. Comput. **31**(5), 392–400 (2013)
25. Zhang, Z.: A flexible new technique for camera calibration. IEEE Trans. Pattern Anal. Mach. Intell. **22**(11), 1330–1334 (2000)

Novel CBCT-MRI Registration Approach for Enhanced Analysis of Temporomandibular Degenerative Joint Disease

Gaëlle Leroux[1,2(✉)], Claudia Mattos[1,3], Jeanne Claret[1,2], Eduardo Caleme[1],
Selene Barone[1], Marcela Gurgel[1], Felicia Miranda[1], Joao Goncalves[4],
Paulo Zupelari Goncalves[1], Marina Morettin Zupelari[1], Larry Wolford[5],
Nina Hsu[1], Antonio Ruellas[6], Jonas Bianchi[7,8], Juan Prieto[8],
and Lucia Cevidanes[1]

[1] University of Michigan, Ann Arbor, USA
`gaellel@umich.edu`
[2] CPE Lyon, Lyon, France
[3] Universidade Federal Fluminense, Niterói, Brazil
[4] Araraquara Dental School, Araraquara, Brazil
[5] Baylor College of Dentistry, Dallas, USA
[6] Universidade Federal do Rio de Janeiro, Rio de Janeiro, Brazil
[7] University of Pacific, Stockton, USA
[8] University of North Carolina, Chapel Hill, USA

Abstract. Temporomandibular Degenerative Joint Disease (TM DJD) is characterized by chronic and progressive degeneration of the joint, leading to functional limitations. Converging on enhancing TM DJD diagnosis, prognosis, and personalized patient care, multimodal Cone Beam Computed Tomography (CBCT) and Magnetic Resonance Imaging (MRI) registration aims to allow comprehensive understanding of the articular disc and subchondral bone alterations towards elucidating the onset, advancement, and progression of TM DJDs. This study proposes a novel multimodal image registration (fusion) approach that combines image processing techniques with mutual information to register MRI to CBCT images, enhancing TMJ complex visualization and analysis. The algorithm leverages automated image orientation, resampling, MRI inversion, normalization and rigid mutual information registration methods to align and overlay multimodal images while preserving anatomical details. Evaluation on 70 CBCT and 70 MRI scans acquired at the same time points for 70 TM DJD patients demonstrates robustness to variations in image quality, anatomical morphology, and acquisition protocols. By integrating MRI soft tissue information with CBCT bony details, this novel open-source tool available in the 3D Slicer platform provides a more comprehensive patient-specific TM DJD model. The current 98.75% success rate, with mean absolute rotation differences of $1.53° \pm 1.75°, 1.69° \pm 1.54°$, and $2.70° \pm 2.89°$ in Pitch, Roll and Yaw

Granted by R01-DE024450.

respectively; and translation differences of 0.92mm ± 1.64mm, 0.98mm ± 0.85mm, and 0.5mm ± 0.43mm in respectively the Left-Right, Antero-Posterior and Supero-Inferior axes. The proposed approach has potential to enhance care for TM DJD and other conditions requiring multimodal images.

Keywords: Multimodal image · Fusion · Degenerative joint disease

1   Introduction

Temporomandibular joint (TMJ) disorders are complex conditions affecting the jaw joint and surrounding tissues [1]. Accurate diagnosis and treatment planning require detailed visualization of both bony structures and soft tissues, including the articular disc, ligaments, and musculature. Cone-beam computed tomography (CBCT) is widely used in dental and maxillofacial imaging due to its high spatial resolution and relatively low radiation dose [2]. However, CBCT has limited soft tissue contrast and cannot adequately visualize the disc and surrounding tissues crucial for TMJ assessment [3].

Magnetic resonance imaging (MRI) provides superior soft tissue delineation without ionizing radiation, making it an ideal complement to CBCT for comprehensive TMJ evaluation [4–6]. Integrating MRI and CBCT data represents a significant advancement in craniofacial assessment, offering unprecedented diagnostic accuracy and treatment planning precision [6,7].

Despite the recognized value of utilizing both modalities, integration remains challenging due to differences in patient positioning, image resolution, and field of view. Traditional registration algorithms are less effective due to the inherent differences in information provided by each modality, necessitating advanced, automated solutions to ensure accurate and efficient data fusion [8]. Manual registration is time-consuming and prone to inter-observer variability [7].

This study aims to address these limitations by developing an algorithm for automated MRI to CBCT registration, enabling efficient and accurate fusion of the complementary information provided by each modality. The novel approach employs image processing methods to achieve robust alignment and natural-looking integration of the multimodal images. By providing a holistic 3D model of the patient's TMJ anatomy [9,10], this technique has the potential to greatly enhance diagnostic capabilities and facilitate personalized treatment strategies.

2   Materials

A total of 70 CBCT and 70 MRI scans of the head in Digital Imaging and Communications in Medicine (DICOM) format were used in this work. The images were acquired at different clinical centers with different scanners, acquisition protocols, and fields of view. All DICOM files were anonymized removing all identifiable personal information using the 3D Slicer Batch Anonymizer module.

The University of Michigan Institutional Review Board (IRB) HUM00239207 waived the requirement for informed consent and granted IRB exemption. The CBCT and MRI scans were acquired using standard clinical protocols without any additional imaging performed for research purposes. All images were anonymized and stripped of protected health information prior to being transferred to the researchers. The data was securely stored on encrypted servers with access restricted to authorized personnel only.

3   Methods

3.1   Data Preprocessing

To achieve accurate registration of MRI to CBCT scans, we developed a novel pipeline that combines both newly developed automated procedures and previously developed tools. The overall workflow of the MR2CBCT registration process is illustrated in Fig. 1. Initially, the CBCT files were not oriented because of the inconsistency in imaging acquisition protocols, patient position during image acquisition, and settings used on scanners in different clinical centers [11]. Therefore, the first step involved orienting and centering these CBCT scans to a common frame of reference. To do this, we used Automated Orientation available on 3DSlicer [12]. Similarly, the MRI data required orientation and centering to align with the CBCT scans.

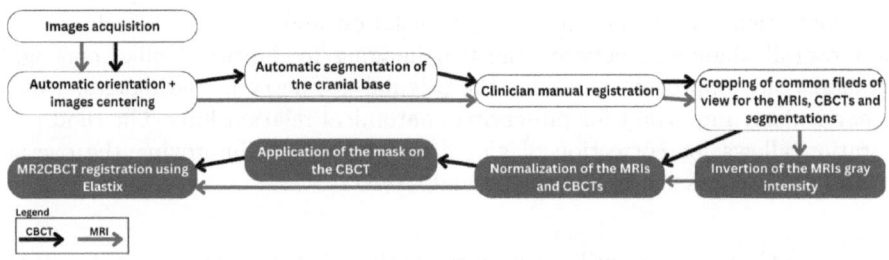

Fig. 1. Workflow of the MR2CBCT registration. The proposed novel pipeline consists of both new automated procedures (shown in green) and previously available tools (shown in gray) that are leveraged to build the overall registration workflow. (Color figure online)

Following this, a clinician expert performed a manual registration using the 3D Slicer Transforms tool to register the CBCT scans with the MRI scans. This manual alignment was essential to establish an initial correspondence between the two modalities. Once the initial registration was achieved, a consistent bounding box was used to crop both the CBCT and MRI scans around the temporomandibular joint (TMJ) area, which was the primary region of interest for our analysis. The machine learning models for CBCT segmentation of the cranial base were automatically computed using the tools called AMASSS [13] from the 3D Slicer 5.6.2 and used as stable regions of reference for registration. The MRI images underwent inversion of the gray level intensity values to better match the contrast of the CBCT scan, as MRI and CBCT images typically present different intensity distributions (Fig. 2). After inversion, we normalized the intensity values of the MRI and CBCT images to a common range, typically [0, 100] for MRI and [0, 75] for CBCT, to ensure consistent intensity scales over the regions of interest in our two images. The preprocessed images were then saved and used in the registration process.

3.2 MRI to CBCT Registration (Fusion)

Following the clinician's manual registration and preprocessing steps, the Elastix mutual information rigid registration method was applied to refine the alignment between the MRI and CBCT scans [14]. The steps involved in this process were:

Mask Application: The mask obtained from the segmentation of the CBCT was applied to the preprocessed CBCT image, isolating the cranial base region to serve as a stable reference for registration with the MRI. This step ensures that the registration focuses on the most relevant anatomical structures and reduces the influence of noise or artifacts in other regions.

Rigid Registration: The rigid registration approach in Elastix optimizes the transformation parameters, including translation and rotation, to achieve the best overall alignment between the two imaging modalities. Unlike non-rigid methods, this approach maintains the original geometry of the images, which is particularly important for preserving anatomical relationships. The rigid registration allows for correction of global misalignments, improving the overall spatial correspondence between MRI and CBCT.

Optimization: Elastix utilizes an optimization algorithm to iteratively adjust the transformation parameters to maximize the mutual information between the images. Mutual information is a statistical measure that quantifies the amount of information obtained about one image given the other, making it suitable for multimodal registration. The optimization process seeks to find the transformation that results in the highest mutual information, indicating the best alignment between the MRI and CBCT scans.

Transformation Parameters: The optimized transformation parameters, including the translation and Euler angles (rotation), were extracted and analyzed to quantify the registration accuracy in six degrees of freedom (DOF). These parameters provide a quantitative measure of the registration performance and can be used to assess the reliability of the registration results.

The Elastix mutual information rigid registration approach leverages the strengths of both imaging modalities, combining the soft tissue contrast of MRI with the bony detail of CBCT. By iteratively optimizing the transformation parameters to maximize mutual information, this method aims to achieve a more precise and robust alignment compared to manual registration alone. The application of the CBCT segmentation mask aims to enhance the registration accuracy by focusing on the most stable and relevant anatomical regions. The goal of this approach is for the resulting registered images to provide a comprehensive 3D model of the TMJ, enabling improved visualization and analysis of both soft tissue and bony structures.

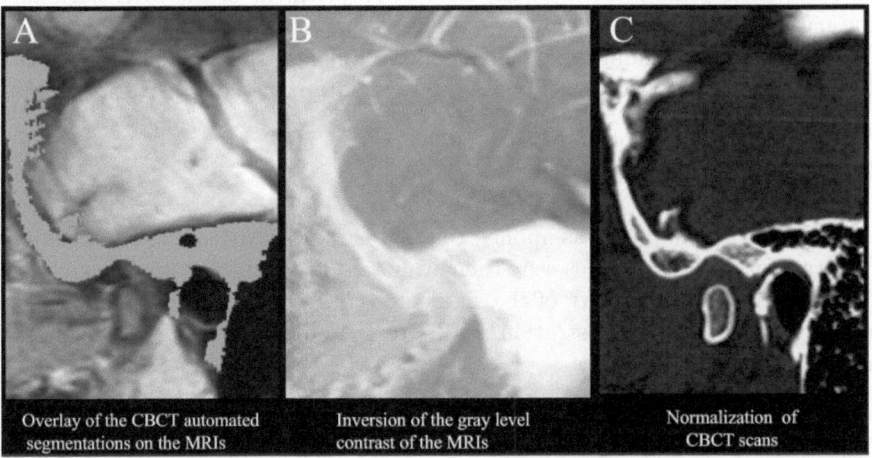

Overlay of the CBCT automated segmentations on the MRIs

Inversion of the gray level contrast of the MRIs

Normalization of CBCT scans

Fig. 2. Preprocessing steps for MRI-CBCT registration. (A) Overlay of the automated CBCT segmentation (shown in green) on the MRI scan, highlighting the cranial base as a stable region of reference. (B) Inversion of the gray level contrast of the MRI scan to better match the intensity characteristics of the CBCT scan, facilitating the registration process. (C) Normalization of the CBCT scan to achieve consistent intensity scales over the regions of interest between the MRI and CBCT images. (Color figure online)

3.3   Evaluation Metrics

To quantify the quality of registration between MRI and CBCT images, a multi-step approach was employed. Initially, a clinician performed manual registration to align the MRI with the CBCT using the 3D Slicer Transforms module, providing a baseline for comparison. We then calculated the registration matrices for the whole sample using the Elastix mutual information rigid registration. The voxel-based registration was quantitatively assessed in the six degrees of freedom of the translation (Left-Right, Antero-Posterior and Supero-Inferior axes) and rotation (Pitch, Roll, and Yaw axes). Summary statistics were computed in Jamovi software version 2.3.28 to report for the differences in each degree of freedom, including the mean difference, mean absolute difference, minimum and maximum absolute differences, and 75th and 90th percentiles of the absolute differences, and graphically display the errors distribution. The quality of voxel-based registration was verified through visual inspection by an expert clinician. The clinician's quality control check focused on the alignment of key anatomical structures and overall spatial correspondence between the two imaging modalities. This quality control allowed detection of clinically relevant registration improvements not captured by the quantitative metrics. Cases were classified as successfully registered if they demonstrated clear visual improvement over the clinician manual registration and if the linear differences in the MR translation were less than 4 mm relative to the CBCT. Adjustment of the MRI normalization parameters was used to improve the precision of the MRI registration to the CBCT. As a gold standard, a panel of expert clinicians reviewed the results after the method was applied, providing a final validation of the registration quality.

4   Results

The registration method was applied to a dataset of 70 MRI-CBCT image pairs. The transformation matrices were then evaluated for quality control by an expert clinician, and the variability of the six degrees of freedom was assessed by comparing the automated registration results to the clinicians' gold standard registration (Table 1 and Fig. 3).

Table 1. Differences in six Degrees of Freedom between clinician and Elastix Image Registration.

ROTATION (°)			
	Pitch	Roll	Yaw
Mean difference (SD*)	0.49 (2.28)	-0.62 (2.21)	-0.99 (3.84)
Mean absolute difference (SD*)	1.53 (1.75)	1.69 (1.54)	2.70 (2.89)
Minimum absolute difference	0.00	0.02	0.03
Maximum absolute difference	8.72	6.79	16.60
75th percentile of absolute difference	1.93	2.56	3.04
90th percentile of absolute difference	3.51	3.32	5.10
TRANSLATION (mm)			
	LR	AP	SI
Mean difference (SD*)	0.3 (1.86)	0.89 (0.94)	-0.20 (0.63)
Mean absolute difference (SD*)	0.92 (1.64)	0.98 (0.85)	0.50 (0.43)
Minimum absolute difference	0.01	0.01	0.00
Maximum absolute difference	13.10	3.52	2.57
75th percentile of absolute difference	1.16	1.48	0.72
90th percentile of absolute difference	1.55	2.12	0.95

* Abbreviation of Standard Deviation (SD)

The results indicate small differences, with mean absolute rotation differences of 1.53°, 1.69° and 2.70°, respectively in Pitch, Roll and Yaw, and mean absolute translation differences of 0.92 mm, 0.98 mm and 0.50 mm respectively in the Left-Right, Antero-Posterior and Supero-Inferior axes. Sixty-nine out of 70 cases presented less than 4 mm in the MRI translation in each axis relative to the CBCT compared to the clinician manual registration. Eleven cases whose MRIs presented darker gray level intensity had a difference of MR translation >2mm and <4mm. Fifty-two cases presented less than 2 mm in the MR translation in each axis relative to the CBCT compared to the clinician manual registration, with improved MRI to CBCT image registration in the clinician quality control check (Fig. 4). The sequence of image processing steps for images centering, orientation, resampling, MRI inversion, normalization and Elastix mutual intensity registration of MRI to CBCT were deployed in Github as a 3D Slicer module for multimodal registration called MR2CBCT.

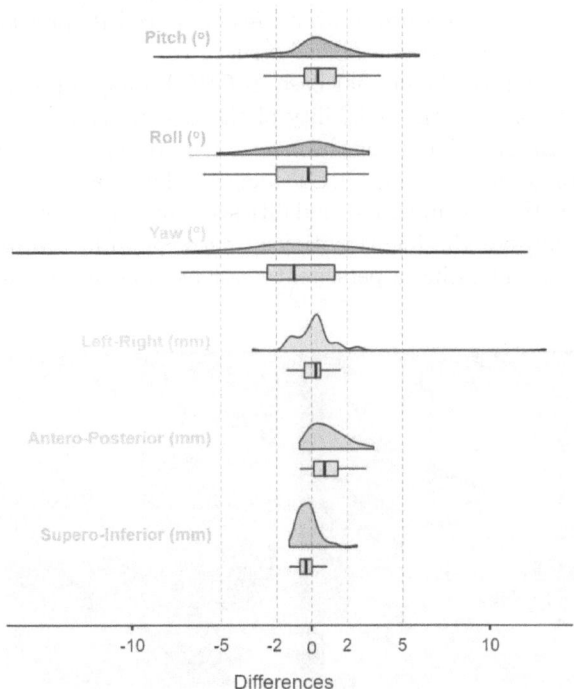

Fig. 3. Box plots with overlaid density plots showing the differences in six degrees of freedom (Pitch, Roll, Yaw, Left-right, Antero-posterior, and Supero-inferior) between the clinician registration and the Elastix registration approaches. The boxes represent the interquartile range (IQR) between the 25th and 75th percentiles, with the median marked by the horizontal line inside the box. The whiskers extend to the most extreme data points within 1.5 times the IQR from the box edges. Outliers beyond the whiskers are plotted as individual points. The density plots on either side of the boxes illustrate the distribution of the data points, with the width of the shaded area representing the proportion of data at each value. Positive and negative values indicate the direction of differences between the clinician and Elastix registration approaches.

5   Discussion

The present study introduces a novel automated method for registering MRI to CBCT scans, focusing on the TMJ region. The proposed approach addresses the limitations of existing methods by integrating image processing techniques to perform robust and accurate multimodal image registration. The results demonstrate the effectiveness of the developed pipeline in aligning MRIs and CBCTs, enabling the fusion of complementary information provided by each modality.

Our study underscores the critical importance of thorough preprocessing in achieving accurate MRI-CBCT registration. The initial steps of orientation, centering, and manual approximation proved fundamental in establishing a common frame of reference, addressing a key challenge in multimodal imaging [11,12]. By utilizing the Automated Orientation tool in 3D Slicer and incorporating clinical expertise, we were able to overcome the inherent differences in image acquisition between MRI and CBCT. The focused cropping of the TMJ area not only streamlined the registration process but also enhanced its precision by concentrating on the most relevant anatomical structures.

The segmentation of the cranial base in CBCT images provided stable reference regions, improving the reliability of the registration process. Our image enhancement techniques, including the inversion of MRI gray level intensities and normalization of both MRI and CBCT images, played a vital role in bridging the gap between the two imaging modalities [7,10]. These preprocessing steps significantly facilitated the identification of corresponding anatomical features across modalities, addressing a persistent challenge in multimodal registration.

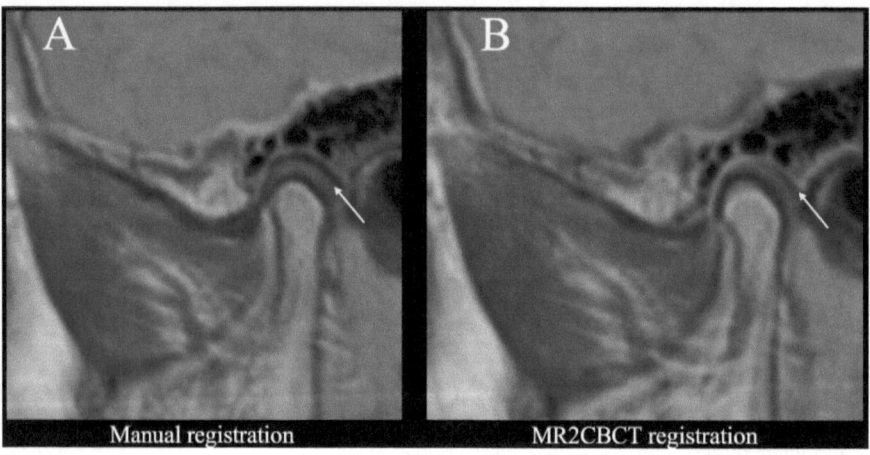

Fig. 4. MRI-CBCT Overlays: (A) Manual registration; (B) Post MR2CBCT Registration using the Cranial Base as a stable region of reference. Note that the CBCT was taken with the mouth slightly open which may have been challenging for the clinician performing manual registration; also note how the fit of the cranial base was improved in the mutual information registration performed by the MR2CBCT algorithm in B.

When considering differences between the gold standard and Elastix registrations, the smallest linear displacement errors were observed in the supero-inferior direction, while the greatest in antero-posterior direction. Regarding rotational differences, the largest error was in pitch. These findings are similar to the directions of greater differences in automated orientation and registration for multiple CBCT scans tested previously [12]. Although these errors can be minimized by future research, they are already within a clinically acceptable range.

Successful integration of MRI and CBCT scans has profound implications for TMJ disorder diagnosis, treatment planning, and research. By providing a holistic 3D model of TMJ anatomy, including both hard and soft tissues, clinicians can gain a more comprehensive understanding of underlying pathologies [3,6]. This enhanced visualization enables identification of subtle changes in articular disc, ligaments, and musculature, invisible on CBCT alone, guiding personalized treatment strategies [8]. The proposed method also has potential to advance TMJ disorder research, opening new opportunities for large-scale studies investigating etiology, progression, and treatment outcomes [1,2].

Despite the insights gained, this study has some limitations. Eleven cases with darker MRI gray level intensity presented a difference of MR translation >2mm and <4mm. Although they were considered successful cases, there is still space for further refinement of preprocessing techniques to handle varied image qualities. Future studies should validate the proposed method on larger, more heterogeneous datasets to assess generalizability. Establishing target points on both MRI and CBCT images or creating MRI segmentations would allow us to incorporate quantitative metrics like Target Registration Error (TRE) and Dice Similarity Coefficient. This future addition will provide a more comprehensive evaluation of registration quality [7].

The proposed automated registration method demonstrates high success rates and has the potential to significantly enhance diagnosis, treatment planning, and research. The comprehensive preprocessing pipeline, combined with the rigid registration approach, enables accurate alignment of MRI and CBCT, providing a holistic 3D model of TMJ anatomy. The proposed pipeline was tested and implemented as functionalities of a free open-source module (available at https://github.com/DCBIA-OrthoLab/SlicerAutomatedDentalTools) with a user-friendly interface in 3DSlicer. Future work should focus on further refinement of preprocessing techniques, quantitative comparisons with various registration approaches as well as incorporate automated segmentation of the articular disc to improve the robustness and clinical applicability of the proposed method.

6   Conclusion

The novel MRI to CBCT registration method developed in this study represents a significant advancement in multimodal image fusion for TMJ disorders. The proposed approach integrates state-of-the-art image processing techniques to enable accurate and efficient alignment of MRI and CBCT scans, providing a comprehensive 3D model of the patient's TMJ anatomy. The results demonstrate a high success rate and small mean absolute differences in rotation and translation, indicating the robustness of the registration approach. This holistic visualization has the potential to enhance diagnostic capabilities, facilitate personalized treatment planning, and advance research in the field of TMJ disorders.

Disclosure of Interests. The authors have no competing interests to declare that are relevant to the content of this article.

References

1. Smith-Bindman, R et al.: Trends in use of medical imaging in US health care systems and in Ontario, Canada, 2000–2016. In: Jama 322.9 (2019), pp. 843–856
2. Kumar, M., et al.: Cone beam computed tomography-know its secrets. J. int. oral health JIOH **7**(2) 64 (2015)
3. Florkow, M.C., et al.: Magnetic resonance imaging versus computed tomography for Three-Dimensional bone imaging of musculoskeletal pathologies: a review. J. Magn. Reson. Imaging **56**(1), 11–34 (2022)
4. Niraj, L.K., et al.: MRI in dentistry-A future towards radiation free imaging-systematic review. J. clin. diagn. res. JCDR **10**(10), ZE14 (2016)
5. Bruno, F., et al.: Advanced magnetic resonance imaging (MRI) of soft tissue tumors: techniques and applications. Radiol. Med. (Torino) **124**, 243–252 (2019)
6. De Schepper, A.M., et al.: Magnetic resonance imaging of soft tissue tumors. Eur. Radiol. **10**, 213–223 (2000)
7. Wang, Y., et al.: Diagnostic efficacy of CBCT, MRI, and CBCT-MRI fused images in distinguishing articular disc calcification from loose body of temporomandibular joint. Clin. Oral Invest. **25**, 1907–1914 (2021)
8. Tai, K., et al.: Preliminary study evaluating the accuracy of MRI images on CBCT images in the field of orthodontics. J. Clin. Pediatr. Dent. **36**(2), 211–218 (2011)
9. Scarfe, W.C, Farman, A.G.: What is cone-beam CT and how does it work? Dent. Clin. North Am. **52**(4), 707– 730 (2008)
10. Al-Saleh, M.A., et al.: MRI alone versus MRI-CBCT registered images to evaluate temporomandibular joint internal derangement. Oral Surg. Oral Med. Oral Pathol. Oral Radiol. **122**(5), 638–645 (2016)
11. Stamatakis, H.C., et al.: Head positioning in a cone beam computed tomography unit and the effect on accuracy of the three-dimensional surface mode. Eur. J. Oral Sci. **127**(1), 72–80 (2019)
12. Anchling, L., et al.: Automated orientation and registration of cone-beam computed tomography scans. In: Workshop on Clinical Image-Based Procedures, pp. 43–58. Springer (2023).https://doi.org/10.1007/978-3-031-45249-9_5
13. Gillot, M., et al.: Automatic multi-anatomical skull structure segmentation of cone-beam computed tomography scans using 3D UNETR. PLoS One **17**(10), e0275033 (2022)
14. Klein, S., et al.: elastix: a toolbox for intensity-based medical image registration. IEEE Trans. Med. Imaging **29**(1), 196–205 (2010). https://doi.org/10.1109/TMI.2009.2035616.

Abdominal Ultrasound Similarity Analysis for Quantitative Longitudinal Liver Fibrosis Staging

Eung-Joo Lee[1], Vivek K. Singh[1], Elham Y. Kalafi[1], Peng Guo[1],
Arinc Ozturk[1], Theodore T. Pierce[1], Brian A. Telfer[2], Anthony E. Samir[1],
and Laura J. Brattain[2]([✉])

[1] Department of Radiology, Massachusetts General Hospital, Boston, MA, USA
[2] MIT Lincoln Laboratory, Lexington, MA, USA
Laura.Brattain@ucf.edu

Abstract. Non-alcoholic fatty liver disease is one of the most common diffuse liver diseases worldwide, affecting approximately 25–30% of the global population. Accurate liver fibrosis staging and longitudinal tracking are crucial for effective care. Core needle biopsy is the current gold standard but it is invasive, inaccurate, and costly. Abdominal B-mode ultrasound (US) in conjunction with shear wave elastography (SWE) offers a non-invasive alternative but suffers from high variability, impairing quantitative assessments. Consistent liver views in the abdominal ultrasound during multiple visits will likely result in consistent SWE-based fibrosis staging. This study presented a proof-of-concept pipeline for identifying the most anatomically similar transverse view in the right liver lobe in abdominal US videos across multiple visits. Our three-stage framework consisted of liver view classification, liver capsule and hepatic vessel segmentation, and quantitative similarity measurement. We used a pretrained EfficientNet-B3 network for liver view classification, achieving 99% accuracy. We then used Efficient-UNet to segment the liver capsule and hepatic vessels, obtaining Dice scores of 90% and 58%, respectively. The classification and segmentation outputs were used for similarity analysis. We evaluated four similarity metrics including deep image structure and texture similarity (DISTS), Root Mean Squared Error (RMSE), Normalized Cross Correlation (NCC), and structural similarity index measure (SSIM), with SSIM resulting in the best results. This pipeline has the potential to improve SWE-based quantitative fibrosis staging and enable cost effective longitudinal tracking.

Keywords: Abdominal liver ultrasound · Deep learning · Longitudinal tracking · Similarity analysis

E.-J. Lee, V. K. Singh, E. Y. Kalafi—Contributed equally to this work
E.-J. Lee, V. K. Singh, L. J. Brattain—As of the submission of the paper, EJL is at the University of Arizona, VKS is at the Barts Cancer Institute, Queen Mary University of London, and LJB is with University of Central Florida College of Medicine.

K. Drechsler et al. (Eds.): CLIP 2024, LNCS 15196, pp. 73–82, 2024.
https://doi.org/10.1007/978-3-031-73083-2_8

1 Introduction

Non-alcoholic fatty liver disease is one of the most common diffuse liver diseases worldwide, affecting approximately 25–30% of the global population [1]. Proper fibrosis staging and longitudinal tracking are crucial for effective care management. Biopsy, the current gold standard, is invasive, costly, and prone to sampling error and operator variability [2]. B-mode ultrasound (US), in conjunction with shear wave elastography (SWE), offers a non-invasive alternative [3, 4]. A consistent US scanning protocol is key to consistent SWE-based liver stiffness measurement [5], but variability in obtaining liver views hampers longitudinal tracking and reduces sensitivity to changes. SWE estimation at similar anatomical locations over multiple visits can reduce SWE measurement variability and track liver fibrosis progression more robustly. Qualitative analysis alone does not guarantee sonographers to obtain consistent measurements at the same locations across exams. An algorithmic solution is needed to address noise in the ultrasound (e.g., speckles, shadows, and low signal-to-noise ratio etc.) and operator and patient variability. Existing metrics like Root Mean Squared Error (RMSE) [6] and Normalized Cross Correlation (NCC) [7] only provide pixel-wise measurements, not considering liver anatomy. Deep learning (DL) methods have shown state-of-the-art results in tasks like object detection, image reconstruction, segmentation, and classification [8–13]. Siamese networks [14], a subset of DL methods, use convolutional neural networks (CNNs) to compare reference and target images for pattern recognition, anomaly detection, landmark tracking, etc. [15–17]. A reference image is an image used as a standard or baseline in image processing tasks. It serves as a point of comparison for other images. A target image is the image that is being analyzed or processed in relation to the reference image. A fully convolutional Siamese network has been reported to identify anatomical landmarks in US images [18], while another estimated disease severity in retinal and knee images [19]. However, these methods are often task-specific and not generalizable to other clinical applications. We propose a novel framework for consistent B-mode liver image acquisition to enable reproducible SWE measurement. This framework includes: 1) liver view classification, 2) liver capsule and hepatic vessel segmentation, and 3) similarity analysis based on the outputs from 1) and 2). We hypothesize that this approach can guide sonographers to the similar liver view in abdominal ultrasound during each visit and obtain consistent SWE measurements. To our knowledge, this is the first attempt to determine the similar anatomical liver right lobe transverse view (LRLTV) across multiple visits. Four similarity metrics, including deep image structure and texture similarity (DISTS) [20], RMSE [6], NCC [7], and structural similarity index measure (SSIM) [21], were implemented and analyzed.

2 Materials and Methods

2.1 Datasets

With Institutional Review Board (IRB) approval, US data were retrospectively collected and consisted of static B-mode images and two-visit longitudinal videos (cine-loops) from two ultrasound vendors/models including Supersonic Imaging (SSI) and GE LOGIQ E9.

DS1: GE Static B-mode. Total of 477 patients (477 images), split into train (70%), validation (10%), and test (20%) sets to train the liver classification and segmentation models. Additionally, 50 patients (370 images) were used to develop the hepatic vessel segmentation method, split into train (35 patients), validation (5 patients), and test (10 patients) sets.

DS2: SSI Static B-mode. Total of 161 patients with 2D B-mode liver US images, each of 8 to 12 US images. We selected 50 patients, each with 4 US images and matching annotations. The data were split into train (60%), validation (10%), and test (30%) sets and were used to fine-tune the liver capsule and hepatic vessel segmentation model.

DS3: GE Cine-loops. Two-visit videos from 100 patients, 1 video per patient. We selected 64 patients with liver right lobe transverse view (LRLTV) for this study, each cine-loop containing 8 to 397 US frames.

2.2 Liver View Similarity Analysis Framework

Our proposed multi-stage similarity framework includes liver view classification, liver capsule, hepatic vessel segmentation, and similarity measurement (Fig. 1). In Stage 1, an EfficientNet-B3 [22] was used to classify liver views in the abdominal US. The primary objective was to exclude images that did not contain quality liver view. A 'quality liver view' is defined as a liver view that contains the key anatomical landmarks such as liver capsule and Morrison's pouch, with minimal shadow artifacts and other interfering factors. The ground-truth was provided by a medical doctor. Images without any liver view were excluded (Fig. 2). Liver view classification is crucial in our proposed framework as Stage 2 depends on a good liver view that includes the presence of the liver capsule and hepatic blood vessels. The pre-trained EfficientNet-B3 [22] used an Inverted Residual Block called MBConv with a squeeze-and-excitation block to extract spatial and global features efficiently.

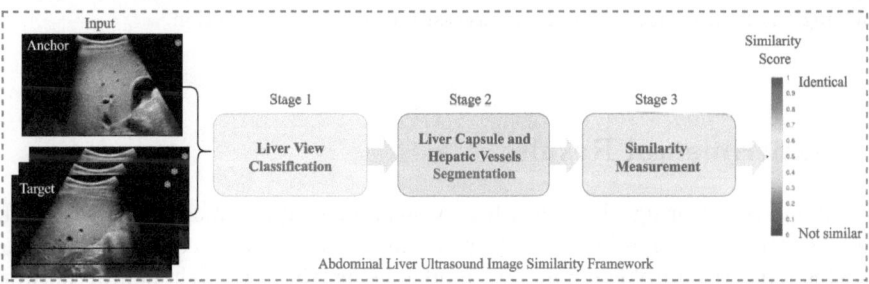

Fig. 1. Proposed framework for liver view similarity analysis in abdominal ultrasound. It consists of three stages: liver view classification, liver capsule, hepatic vessel segmentation, and similarity measurement.

In Stage 2, an Efficient-UNet [23] model was chosen to segment the liver and hepatic vessels because it is computationally efficient. Efficient-UNet has an encoder-decoder network where the encoder uses pre-trained EfficientNet-B3 to extract features like liver textural patterns and edges, while the decoder unpools feature maps with transposed convolutional layers. Skip connections with element-wise feature concatenation enhance segmentation results. The liver and hepatic vessel segmentation models were first developed on DS1 and then fine-tuned with DS2 for generalizability. A weighted combination of binary cross-entropy (BCE) and dice losses were used for the training and validation.

In Stage 3, the liver and vessel masks generated in Stage 2 were used for image similarity analysis. The main reason for using binary masks was to eliminate speckle noise, background artifacts, and machine-generated symbols in the B-mode US, which could interfere with the performance of similarity metrics. DISTS is a deep learning based similarity analysis method. We compared it against conventional RMSE, NCC, and SSIM methods.

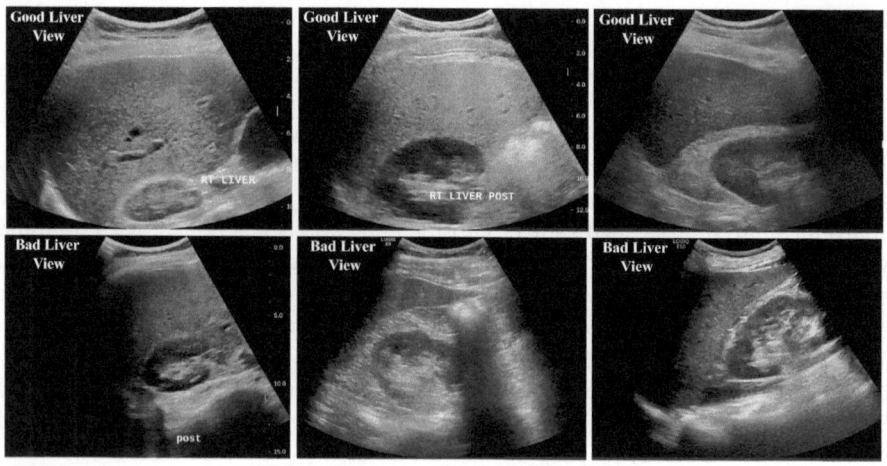

Fig. 2. Example images of good and poor quality liver views in the abdominal US showing the high variability and the necessity of liver view selection.

3 Experimental Results

The similarity scoring algorithm has two-channel inputs including liver masks and hepatic vessel masks. The deep learning model architecture is based on DISTS [20]. We implemented our model using the PyTorch neural network library on a 3.4 GHz Intel Core-i9 with 32 GB of RAM and NVIDIA-RTX 2080Ti GPU with a memory of 11 GB. For proof-of-concept, we applied the model to all the frames in an abdominal ultrasound video with manually selected anchor frame and compared the similarity scores across all other frames in the video.

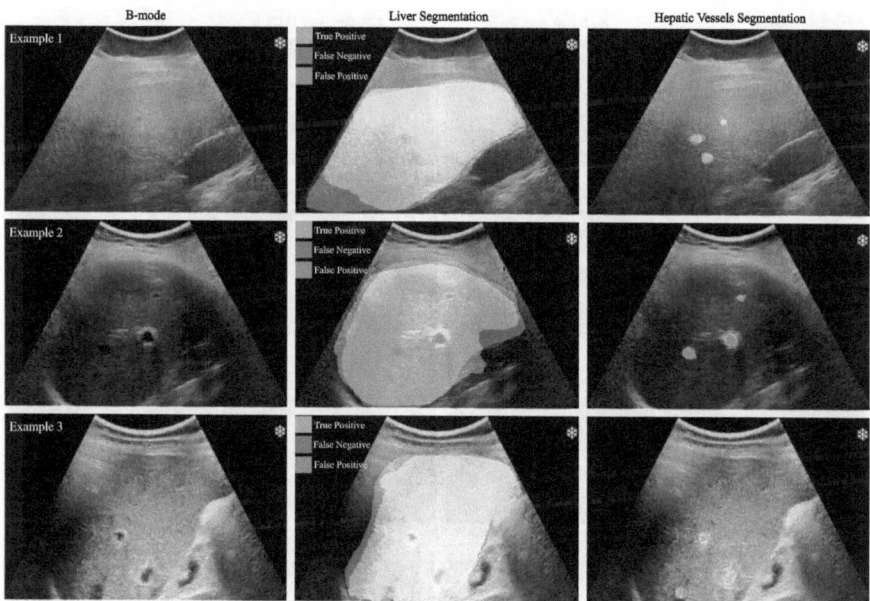

Fig. 3. Three examples of raw B-mode image, liver segmentation, and hepatic vessel segmentation, showing true positives in yellow, false positives in green, and false negatives in red. We observe excellent agreement between the truth and algorithm results. (Color figure online)

3.1 Liver View Classification Results

In Stage 1, DS1 was used for training and evaluation. Data augmentation techniques, such as rotation and horizontal flipping, were applied. The model was trained with an input size of 384 × 384, using the ADAM optimizer with a learning rate of 0.0002 and $\beta_1 = 0.5$, $\beta_2 = 0.999$. The mini-batch size was two, with 100 epochs, using binary cross-entropy loss. We compared the EfficientNet-B3 liver view classification results with EfficientNet-B0, EfficientNet-B1, and EfficientNet-B2. Our experimental findings showed that, on the test set, EfficientNet-B0, EfficientNet-B1, and EfficientNet-B2 achieved an accuracy score of 94%, 96.11%, and 97.47%, respectively, while EfficientNet-B3 achieved the highest overall accuracy of 99%. From the results, except for one sample, the rest of the images were correctly predicted by the classification model.

3.2 Liver Capsule and Hepatic Vessel Segmentation Results

Efficient-UNet was adapted to this segmentation task. The input image size was 384 × 384 pixels, with binary masks of corresponding liver and vessel images as truth. Pixel values were normalized from 0–255 to 0–1. The optimizer settings were the same as in Stage 1, with a mini-batch size of 4. The model was trained

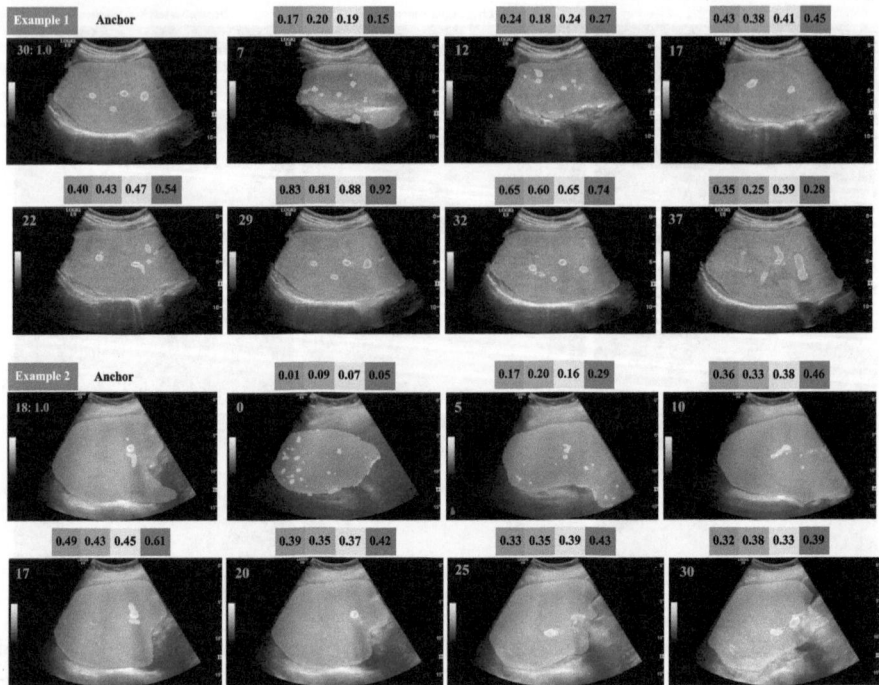

Fig. 4. Examples of similarity scores of four metrics: DISTS, RMSE, NCC, and SSIM. In each example, an anchor image is compared with the target samples. Colors blue, orange, pale green, and red refer to DISTS, RMSE, NCC, and SSIM similarity scores, respectively. We can see qualitatively and quantitatively that in Example 1, Frame 29 is the best match to the anchor image while in Example 2, Frame 17 is the best match. This is consistent with the expectation that in each video, frames closer to the anchor frame resemble the anchor frame more. (Color figure online)

for 100 epochs, saving the best weights based on the highest Dice coefficient score on the validation set. A threshold of 0.5 was used to generate the binary segmentation map of the liver capsule and hepatic vessels. The model achieved Dice scores of 90% on the liver capsule and 58% for hepatic vessel segmentation. We observed that accurate vessel segmentation was challenging due to the small and varying vessel sizes and neighboring shadows. The proposed network segmented clearly identified vessels but failed to delineate the blurry or suspicious boundaries. Figure 3 presents example results of the liver capsule and vessel segmentation with true positive in yellow, false positive in green, and false negative in red. From the visual inspection, it can be seen that Efficient-UNet segmented the liver and vessel region efficiently. However, the model performance degraded in the presence of acoustic shadows.

3.3   Similarity Score Analysis

To compare the performance of similarity scoring, we evaluated four image similarity metrics (i.e., DISTS, SSIM, RMSE, and NCC) on 64 patients. We first manually selected the anchor frame of the best liver view from each video. The rest of the frames in each video served as the target frames. We then applied similarity scoring between the anchor frame and each of the target frames. Since the ultrasound video is of at least 30 frames per second, the frames closer to the anchor frame were expected to have higher similarity.

Figure 4 shows two examples of similarity scoring results on cine-loops generated by various metrics. In Example 1, we chose Frame 30 as the anchor frame and compared it against the rest of the frames (target). It can be seen that Frame 29 achieved the highest similarity scores of 0.83, 0.81, 0.88, and 0.92 by the DISTS, RMSE, NCC, and SSIM, respectively. Upon the visual inspection,

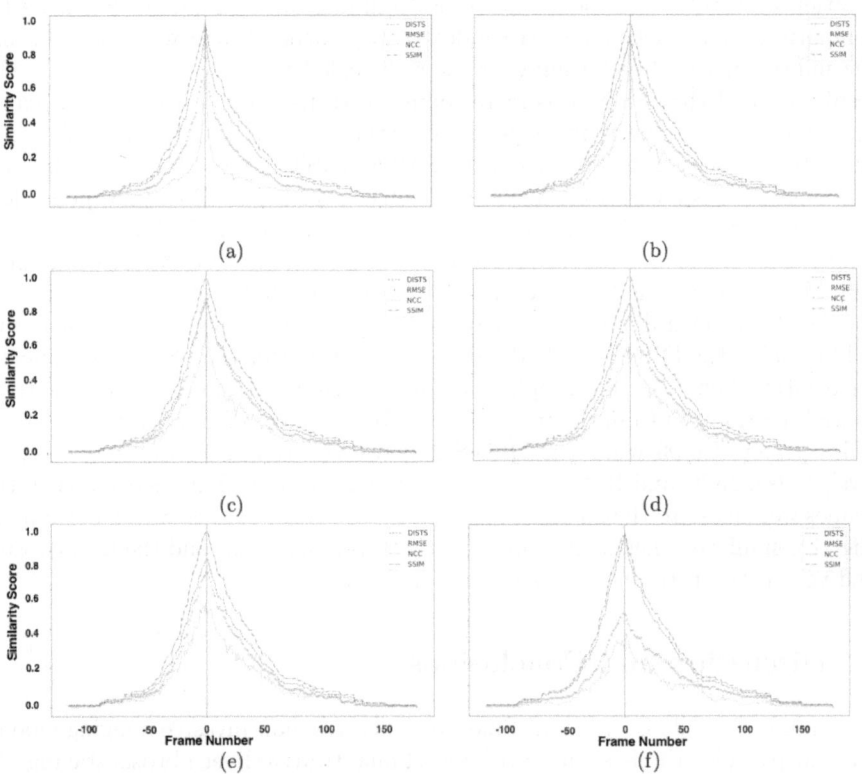

Fig. 5. Proof-of-concept similarity analysis on liver ultrasound videos. Corresponding sub figures of a, b, c, d, e, and f are representative results on anchor frame without variation, with rotation of 15°C, 30°C, 60°C, horizontal flipping, and wrapping, respectively. Four metrics were compared. The SSIM performed the best against other metrics and showed that its scores are much more robust to geometric transforms.

anatomical landmarks, such as the liver capsule and hepatic vessels, represented similar shapes and positions. The segmentation model accurately identified relevant landmarks in anchor and target images. As expected, similarity scores decreased with increasing distance from the anchor frame. Similarly, in Example 2, we selected Frame 18 as the anchor. The neighboring Frame 17 obtained the highest similarity score of 0.61 by SSIM when compared to the other frames. This proof-of-concept study demonstrated the utility of our similarity analysis pipeline.

In clinical settings, liver US scanning is subjected to various external factors such as patient breathing and probe misalignment. Thus, it is necessary to understand the effectiveness of each computed metric on various geometric changes on anchor images, such as translation and rotation. Figure 5 presented the average similarity score plots under various spatial transformations on the 64-patient data above. We computed the average similarity scores between anchor and target frames in conditions including anchor image without variation, with rotation of 15°C, 30°C, 60°C, horizontal flipping, and wrapping. For a better comparison of the 64 videos, the index of the anchor frame was set to 0, and the indices of the target frames (n) were rescaled to $[(-n, 0) \cup (0,+n)]$. The results showed that SSIM was more robust to translation and rotation because it evaluated the similarity in shape and structure, capturing the structural details that are robust to geometric changes. RMSE, DISTS, and NCC methods however showed less consistency under geometric transforms. RMSE, which solely measured pixel-wise differences, was less informative when dealing with binary masks since it did not capture structural changes accurately. NCC, while more suitable for comparing binary data than RMSE, still fell short in accounting for the subtle structural variations that could occur due to geometric transforms. Additionally, the DISTS method did not perform better than SSIM, specifically for rotational changes and flipping operations, likely because it failed to extract spatial information from the masks alone. For instance, with a rotation of 60°C, SSIM achieved 0.95 while the DISTS and RMSE around 0.7 and 0.65, respectively. Also, NCC and RMSE showed more sensitivity to changes even when the frames were close to the anchor frame. Thus, for binary masks, SSIM was a more effective similarity metric in comparing the mask alignment and the liver capsule and vessel structure under geometric transforms.

4   Discussion and Conclusions

Abdominal B-mode US is a real-time, cost effective, non-invasive imaging modality that provides the basis for SWE-based quantitative liver fibrosis staging. To reduce the variability in obtaining the optimal liver view during patient returning visits, we developed a proof-of-concept deep learning-based framework to compute similarity scores between an optimal liver view of LRLTV and a target image frame. Our approach consists of three stages: liver view classification, liver capsule and hepatic vessel segmentation, and similarity measurement. Four image similarity metrics including DISTS, SSIM, RMSE, and NCC were implemented, and their performance was evaluated. The experimental results demonstrated that the segmentation models accurately identified the relevant landmarks in both the anchor and target images. The preliminary similarity results also indicated that with masks as inputs, DISTS, RMSE, and NCC are sensitive to translation and rotation changes, even when the frames are close to the anchor frame. SSIM is more robust to rotation variations and invariant to translations. One limitation of this study is the small datasets for each stage. The next step is to generate ground truth from cine-loops by domain experts and retrain the deep learning models. Future datasets will be expanded to cover the diverse physical characteristics of patients and image variations from ultrasound vendors. Motion artifacts from patient movements will also need to be addressed. We plan to conduct experiments encompassing a wide range of scenarios representing real-world clinical conditions. This will enable us to capture variations in the anchor images and refine the similarity metrics. We hypothesize that liver B-mode similarity analysis can reduce the variability in SWE measurements thus improving liver fibrosis staging. To fully test this hypothesis, in the future, we will design prospective experiments where we will collect abdominal liver US and SWE measurements from a diverse patient population and compare SWE performance with and without the use of similarity analysis. The pipeline presented has the potential to improve SWE-based quantitative fibrosis staging and enable cost effective longitudinal tracking.

Acknowledgments. This material was based upon work supported by the National Institute of Diabetes and Digestive and Kidney Diseases (NIDDK) 5R01DK119860-03. Any opinions, findings, conclusions, or recommendations expressed in this material are those of the author(s) and do not necessarily reflect the views of NIH. Massachusetts General Hospital supported MIT Lincoln Laboratory through Air Force Contract No. FA8702-15-D-0001. The authors wish to thank the following individuals for their clinical assistance: Qian Li, Marian Martin, Firouzeh Heidari, Katie Pope, Hannah Edenbaum, Madhangi Parameswaran, and Siddhi Hegde. ©2024 Massachusetts General Hospital and Massachusetts Institute of Technology.

Disclosure of Interests. TTP was supported by the 2020 American Roentgen Ray Scholar award. He discloses research support from General Electric, the US Department of Defense, Food and Drug Administration, and the National Institutes of Health; royalties from Elsevier Inc.; equity and consultancy for AutonomUS Medical Technologies Inc.; and honoraria from the Massachusetts Society of Radiologic Technologists and Zhejiang Medical Association. AES has provided consulting for numerous healthcare organizations. In the past two years, these include General Electric, Resolve Stroke, Cryosa, Ochre Bio, Rhino Healthtech, and Gerson Lehman Group; is a member of advisory or scientific boards for General Electric, Rhino Healthtech, Ochre Bio, Resolve Stroke, and FNIH; has received research support from Canon, Echosens, General Electric HealthCare, Philips, Siemens, and Supersonic Imagine/Hologic, has received research funding from Analogic Corporation, the US Department of Defense, Fujifilm Healthcare, FNIH, NIH, and General Electric HealthCare; and holds stock options or equity in AutonomUS Medical Technologies, Inc., Evidence Based Psychology LLC, Klea LLC, Katharos Labs LLC, Quantix Bio LLC, Rhino Healthtech, Inc,

Ochre Bio Inc., and Resolve Stroke S.A. The rest of the authors have no competing interests to declare that are relevant to the content of this article.

References

1. Monelli, F., et al.: Cancers **13**, 2246 (2021)
2. Taouli, B., Alves, F.C.: Abdominal Radiology **45**, 3381–3385 (2020)
3. Lee, S.M., et al.: PLoS ONE **12**, e0177264 (2017)
4. Fu, J., Wu, B., Wu, H., Lin, F., Deng, W.: BMC Med. Imaging **20**, 1–9 (2020)
5. Srinivasa Babu, A., et al.: Radiographics **36**, 1987–2006 (2016)
6. Chai, T., Draxler, R.R.: Geoscientific Model Development Discussions **7**, 1525–1534 (2014)
7. Zhao, F., Huang, Q., Gao, W.: Image matching by normalized cross-correlation. In: 2006 IEEE International Conference on Acoustics Speech and Signal Processing Proceedings, vol. 2, pp. II–II. IEEE (2006)
8. Punn, N.S., Patel, B., Banerjee, I.: Abdominal Radiology **49**, 69–80 (2024)
9. Kamiyama, N., Sugimoto, K., Nakahara, R., Kakegawa, T., Itoi, T.: J. Med. Ultrason. **51**, 83–93 (2024)
10. Conze, P.H., Andrade-Miranda, G., Singh, V.K., Jaouen, V., Visvikis, D.: IEEE Transactions on Radiation and Plasma Medical Sciences **7**, 545–569 (2023)
11. Alom, M.Z., et al.: Electronics **8**, 292 (2019)
12. Alzubaidi, L., et al.: Journal of big Data **8**, 1–74 (2021)
13. Guo, Y., Liu, Y., Oerlemans, A., Lao, S., Wu, S., Lew, M.S.: Neurocomputing **187**, 27–48 (2016)
14. Koch, G., Zemel, R., Salakhutdinov, R.: Siamese neural networks for one-shot image recognition. In: ICML Deep Learning Workshop, vol. 2 (Lille) (2015)
15. Nandy, A., Haldar, S., Banerjee, S., Mitra, S.: A survey on applications of siamese neural networks in computer vision. In: 2020 International Conference for Emerging Technology (INCET), pp. 1–5. IEEE (2020)
16. Chen, Z., Zhong, B., Li, G., Zhang, S., Ji, R.: Siamese box adaptive network for visual tracking. In: Proceedings of the IEEE/CVF Conference on Computer Vision and Pattern Recognition, pp. 6668–6677 (2020)
17. Zhang, Z., Peng, H.: Deeper and wider siamese networks for real-time visual tracking. In: Proceedings of the IEEE/CVF Conference on Computer Vision and Pattern Recognition, pp. 4591–4600 (2019)
18. Gomariz, A., Li, W., Ozkan, E., Tanner, C., Goksel, O.: Siamese networks with location prior for landmark tracking in liver ultrasound sequences. In: 2019 IEEE 16th International Symposium on Biomedical Imaging (ISBI 2019), pp. 1757–1760. IEEE (2019)
19. Li, M.D., et al.: NPJ digital medicine **3**, 48 (2020)
20. Ding, K., Ma, K., Wang, S., Simoncelli, E.P.: (2020). arXiv preprint http://arxiv.org/abs/2004.07728
21. Wang, Z., Bovik, A.C., Sheikh, H.R., Simoncelli, E.P.: IEEE Transactions Image Processing **13**, 600–612 (2004)
22. Tan, M., Le, Q.: EfficientNet: rethinking model scaling for convolutional neural networks. In: International Conference on Machine Learning (PMLR), pp. 6105–6114 (2019)
23. Baheti, B., Innani, S., Gajre, S., Talbar, S.: Eff-UNet: a novel architecture for semantic segmentation in unstructured environment. In: Proceedings of the IEEE/CVF Conference on Computer Vision and Pattern Recognition Workshops, pp. 358–359 (2020)

Real-Time Device Detection with Rotated Bounding Boxes and Its Clinical Application

YingLiang Ma[1]([⊠]), Sandra Howell[2], Aldo Rinaldi[2], Tarv Dhanjal[3], and Kawal S. Rhode[2]

[1] School of Computing Sciences, University of East Anglia, Norwich, UK
yingliang.ma@uea.ac.uk
[2] School of Biomedical Engineering and Imaging Sciences, King's College London, London, UK
[3] Warwick Medical School, The University of Warwick, Coventry, UK

Abstract. Interventional devices and insertable imaging devices such as trans-esophageal echo (TOE) probes are routinely used in minimally invasive cardio-vascular procedures. Detecting their positions and orientations in X-ray fluoro-scopic images is important for many clinical applications. Nearly all interventional devices used in cardiovascular procedures contain a wire or wires and are inserted into major blood vessels. In this paper, novel attention mechanisms were designed to guide a convolution neural network (CNN) model to the areas of wires in X-ray images. The first attention mechanism was achieved by using multi-scale Gaussian derivative filters in the first convolutional layer inside the proposed CNN back-bone. By combining these multi-scale Gaussian derivative filters together, they can provide a global attention on the wire-like or tube-like structures. Furthermore, the dot-product based attention layer was used to calculate the similarity between the random filter output and the output from the Gaussian derivative filters, which further enhances the attention on the wire-like or tube-like structures. By using both attention mechanisms, a high-performance CNN backbone was created, and it can be plugged into light-weighted CNN models for multiple object detection. An accuracy of 0.88 ± 0.04 was achieved for detecting an echo probe in X-ray images at 58 FPS, which was measured by intersection-over-union (IoU). Based on the detected pose of the echo probe, 3D echo can be fused with live X-ray images to provide a hybrid guidance solution. Codes are available at https://github.com/YingLiangMa/AttWire.

Keywords: Rotated Object Detection · X-ray Imaging · Attention CNN

1 Introduction

Minimally invasive cardiovascular procedures are routinely carried out to treat diseases such as coronary heart diseases, valvular heart disease, congenital heart diseases and more. The procedure is usually guided using X-ray fluoroscopy and interventional devices and insertable imaging devices are routinely used during the procedure. Real-time object detection for medical devices is one of the most important tasks in hybrid

© The Author(s), under exclusive license to Springer Nature Switzerland AG 2024
K. Drechsler et al. (Eds.): CLIP 2024, LNCS 15196, pp. 83–93, 2024.
https://doi.org/10.1007/978-3-031-73083-2_9

guidance systems as well as robotic procedure systems. Hybrid guidance systems for minimally invasive cardiovascular procedures involves fusing information from Magnetic Resonance Imaging (MRI) images, CT images or real-time 3D transesophageal echo (TOE) with X-ray fluoroscopy [1][2]. Device detection can facilitate the hybrid guidance system using both 3D echo volumes and X-ray fluoroscopic images, and the real-time registration is achieved by detecting the pose of the TOE probe in X-ray images [3]. Real-time device detection also facilitates motion compensation and automatic registration in MRI or CT based hybrid guidance systems [4]. Furthermore, knowing the locations of devices may allow procedures with complete or shared autonomy with robots in the near future.

The detection of the TOE probe and interventional devices in X-ray images has been previously studied. Existing methods can be divided into two categories: traditional computer vision techniques [5][6] and learning-based methods [7–10]. Methods based on traditional computer vision techniques are prone to errors due to image artifacts and the presence of other similar objects. Although learning-based methods have demonstrated a great potential to detect devices robustly, they relied on manual feature selection. Therefore, these methods are not easily transferred to other target devices or detect multiple devices at the same time.

In recent years, state-of-the-art multiple object detection methods have been developed to detect and identify common objects (e.g. vehicles, people, animals and more) [11]. The majority of these methods use axis-aligned bounding boxes to locate the target objects. However, our proposed method requires rotated bounding boxes as medical devices in X-ray images often have arbitrary orientations and rotated bounding boxes are more accurate to determine their locations. Furthermore, applications such as the hybrid guidance using echo and X-ray images requires the orientation of the TOE probe in X-ray images. Few deep-learning based object detection methods using rotated bounding boxes have been developed and they are mainly in the domain of satellite image analysis [12]. However, all existing methods do not meet our requirements for device detection. First, existing methods do not optimize for grayscale X-ray images and lack attention mechanisms for our target objects. Secondly, existing methods do not have sufficient accuracy, robustness or speed to be used in our applications. Therefore, we designed a convolution neural network (CNN) from scratch to achieve our requirements and also to take advantage of additional information available in X-ray images. Many interventional devices contain a wire and wire mesh and are inserted into major blood vessels. Insertable imaging devices are tube-like structures. Therefore, novel attention mechanisms using trainable pre-defined filters and an attention layer were designed to guide our CNN models to the areas of wires in the X-ray images.

2 Method

2.1 Image Acquisition and Image Synthesis

10,072 X-ray images were acquired in 43 different clinical cases using a mono-plane X-ray system at St. Thomas Hospital London. There were 6,533 images from 9 transcatheter aortic valve replacement (TAVR) procedures, 250 images from one atrial fibrillation (AF) procedure guided by X-ray and transesophageal echo images and 3,289 images from 33 standard AF procedures.

As 4,789 images out of total 10,072 images do not contain the TOE probe, a method of image synthesis has been developed to automatically insert an image patch of a TOE probe. It is based on Poisson image editing (PIE) [13][14], which blends an image patch into the context of a destination image. The blending was achieved via solving the Eq. (1).

$$\min_{f_{in}} \iint_{\Omega} |\nabla f_{in} - v|^2 \text{with} f_{in}|_{\partial\Omega} = f_{out}|_{\partial\Omega} \qquad (1)$$

where ∇ is the gradient operator. The goal of Eq. (1) is to find the intensity values f_{in} within the masked area (Ω) of image patch matching with the surrounding values f_{out} of the destination image. A binary mask will be used to create the masked area (Ω), which is the loose selection of the blending object. $\partial\Omega$ is the border of the masked area and v is the image gradient within the masked area. Figure 1 gives an example of PIE.

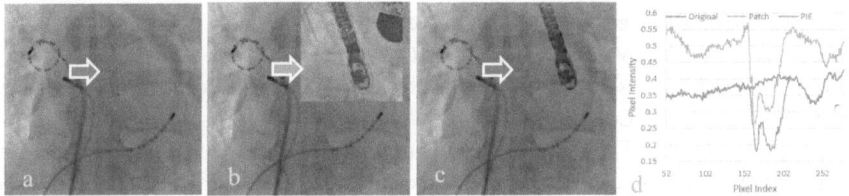

Fig. 1. An example of PIE. (a) The original image. (b) Overlay the image patch with the original image. The red contour is the border of the masked area ($\partial\Omega$). (c) Image after applying PIE. (d) The intensity profiles. Arrows in the images indicate the location of the intensity profiles.

40 image patches were extracted from 40 image sequences which contain the TOE probe. Data augmentation techniques were used to increase the variation of the pose of the TOE probe in X-ray images. Random rotations and translations were applied to the extracted image patches. There are restrictions applied to random rotations and translations to ensure the generated image are anatomically correct.

2.2 The Attention Backbone

Our clinical applications require real-time object detection while maintaining high accuracy and robustness. To achieve this goal, attention mechanisms were designed to take advantage of additional information about the location and structure of medical devices. Many devices contain a wire and wire mesh or tube-like structures. To guide the attention of our CNN models, the multi-scale Gaussian derivative filters were used in the first convolution layer to enhance the visibility of wire-like or tube-like objects [15]. This process involves the calculation of a 2x2 Hessian matrix, and it is computed at every image pixel [16]. The Hessian matrix H consists of second order derivatives that contain information about the local curvature. H is defined such as:

$$H = \begin{bmatrix} L_{xx}(x, y; s) & L_{xy}(x, y; s) \\ L_{yx}(x, y; s) & L_{yy}(x, y; s) \end{bmatrix} \qquad (3)$$

where $L_{xy}(x, y; s) = \frac{\partial^2 L(x,y;s)}{\partial x \partial y}$ and the other terms are defined similarly. Here, $L_{xy}(x, y; s) = L_{yx}(x, y; s)$. $L(x, y; s)$ is an image smoothed by a Gaussian filter of the appropriate scale s. $L(x, y; s)$ is computed as $L(x, y; s) = L(x, y) * G(x, y; s)$, where $*$ is the convolution operator and the Gaussian filter $G(x, y; s) = \frac{1}{2\pi s} e^{-(x^2+y^2)/2s}$. Therefore, Eq. (3) can be converted to

$$ H = \begin{bmatrix} L(x, y) * G_{xx}(x, y; s) & L(x, y) * G_{xy}(x, y; s) \\ L(x, y) * G_{yx}(x, y; s) & L(x, y) * G_{yy}(x, y; s) \end{bmatrix} \tag{4} $$

where $G_{xx}(x, y; s)$, $G_{yy}(x, y; s)$ and $G_{xy}(x, y; s)$ are Gaussian derivatives and are often known as Laplacian of Gaussians (LoG). In practice, we just pre-compute the masks of these Gaussian derivatives, convolve with the input image. By combining these multi-scale Gaussian derivative filters together, they can provide a global attention on the wire-like or tube-like structures.

The architecture of the attention backbone is illustrated in Figs. 3 and Fig. 4 LoG filters are used in the first convolution layer to provide the first attention mechanism. Among 15 filters, there are five groups, and each group contains three LoG filters with the same scale factor s, which are defined as $G_{xx}(x, y; s)$, $G_{yy}(x, y; s)$ or $G_{xy}(x, y; s)$. To accommodate different sizes of objects on the wires, five different scale factors were used in five groups of LoG filters. To calculate the scale factor s_0 for object size r_0, we use $s_0 = ((r_0 - 1)/3)^2$. This equation is motivated by the "3σ" ($s_0 = \sigma^2$) rule that 99% of energy of the Gaussian is within three standard deviations. The final multiscale s_0 is in the range of $0.11 \leq s_0 \leq 9$ and it is based object size from 2 to 10 (Unit is in image pixels) in an image with a 200x200 resolution. The second attention mechanism is achieved by a dot-product based attention layer [17], which calculates the similarity between the random filter output and the output from LoG filters (Fig. 3). The attention layer further enhances the attention on the wire-like or tube-like structures.

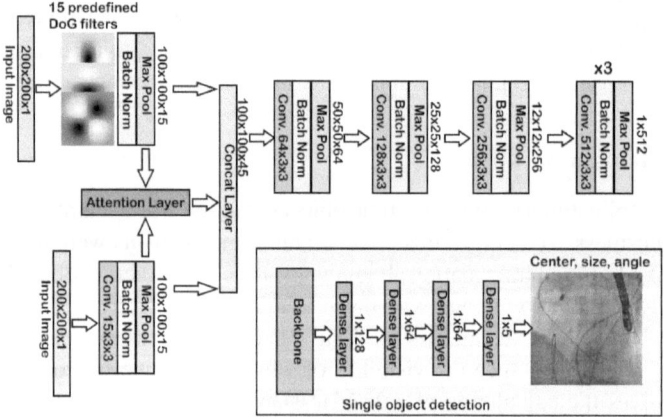

Fig. 2. Attention backbone and an example of its usage in single object detection.

2.3 Single Object Detection

Firstly, a customized CNN was designed for single object detection using the proposed attention backbone. As shown in Fig. 2, the localization of a rotated bounding box is achieved by the output of the final dense layer, which provides five parameters: center (x, y), size (w, h) and angle (δ).

Two object detectors were trained, one for the TOE probe and the other one for the transcatheter aortic valve (before deployment). A modulated rotation loss function was designed and adapted from [18]. It minimizes the difference between the predicted values and ground truth values. All five parameters which define the rotated bonding box were normalized between 0 and 1 to avoid errors caused by objects on different scales. The loss function is defined as:

$$l_{cp} = |x_g - x_p|/W_{img} + |y_g - y_p|/H_{img} \tag{5}$$

$$l_{mr} = min\begin{cases} l_{cp} + |w_g - w_p|/W_{img} + |h_g - h_p|/H_{img} + |\delta_g - \delta_p|/90° \\ l_{cp} + |w_g - h_p|/W_{img} + |h_g - w_p|/H_{img} + |90° - |\delta_g - \delta_p||/90° \end{cases} \tag{6}$$

where l_{cp} is the central point loss, (x_p, y_p) is the predicted center point and (x_g, y_g) is the ground-truth center point. Equation (6) is for the exchangeability of height and width.

As shown in Fig. 3, the activation maps of selected layers were visualized to illustrate the model attentions in the aortic valve detector. The model global attention is clearly on the wires in the first layer and then enhanced by the attention layer. Finally, the model shifts the attention to the local areas of the target object (the aortic valve) in the final convolution layer.

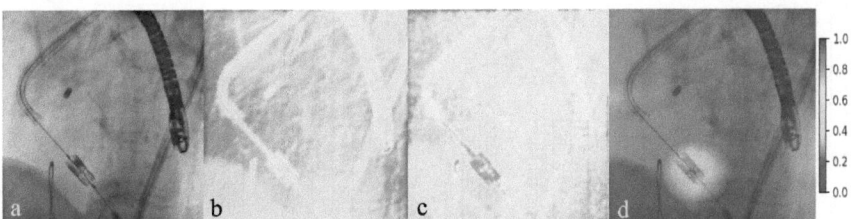

Fig. 3. The original image. (b) The activation map from the convolution layer with 15 LoG filters. (c) The activation map from the attention layer. (d) The activation map from the last convolution layer in the attention backbone(a).

2.4 Multiple Object Detection

Inspired by the CenterNet [19], a one-stage multiple-object detector was designed by using a similar attention backbone proposed in this paper. The one-stage detector can achieve a higher inference speed and it is suitable for real-time applications. The proposed detector is a light-weight CNN model and only contains 3.7M trainable parameters for two-class object detector. As shown in Fig. 4, the proposed CNN model consists of down-sampling layers and up-sampling layers. The model has four outputs. The first one is the center-point heatmap and it is used to localize the center point (x, y) of the rotated bounding box. The second output is used to determine the object size (w, h). The third output is the rotate angle (δ) of the bounding box. The fourth output is the offset output, and it is used to recover from the discretization error caused due to the downsampling of the input. For example, in our model, the input image resolution is 256x256 and the image resolution after the last up-sampling layer is 128x128. If the ground truth of center point is (x_g, y_g) in center heatmap output, the corresponding ground truth of center point in the input image is $(2x_g + \varepsilon_x, 2y_g + \varepsilon_y)$. Both ε_x and ε_y are discretization errors and they are either 0 or 1 in our model.

The proposed CNN model not only outputs a rotated bounding box for each object but also outputs a confidence value. Therefore, the model can predict whether the target object exists in the image or not. The model also can achieve multiple object detection as it has multiple channels of center heatmaps and each channel can localize the center points of one class of objects. The ground-truth heatmap for center points is not defined as either 0 or 1 because locations near the target point should get less penalization than locations far away. Therefore, Gaussian heatmap $e^{\frac{P-P_g^2}{2\sigma^2}}$ was used and P is the predicted center point and P_g is the ground truth. σ is set to 1/3 of the radius, which is determined by the size of objects. Focal loss [20] is used in the output for center-point heatmap and it is mainly to solve the problem of imbalanced classification in target detection. The loss functions for the remaining outputs are L1 loss function. Figure 5 presents some results of center-point heatmaps and detection results.

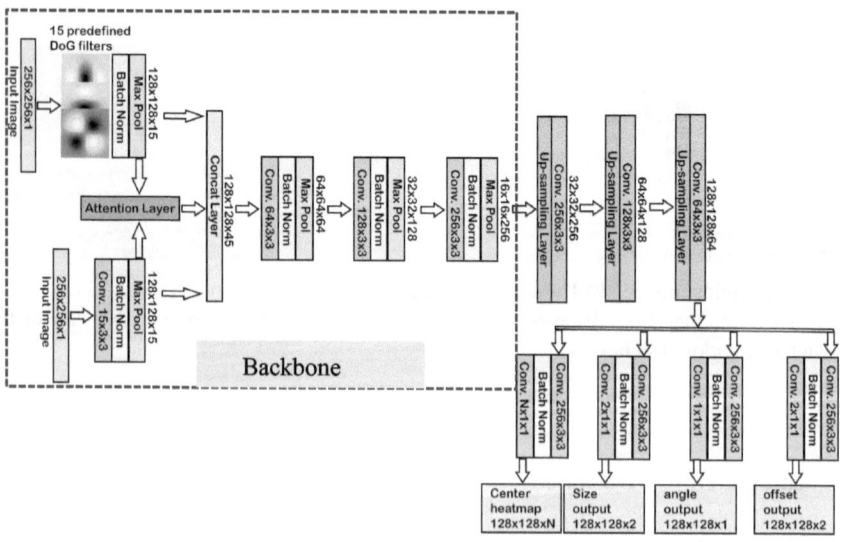

Fig. 4. The CNN architecture for multiple object detection. N is the number of classes.

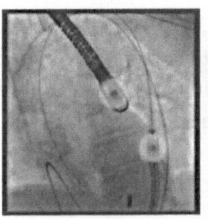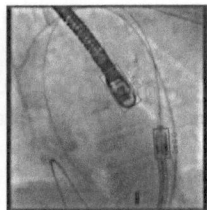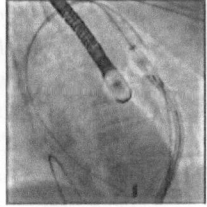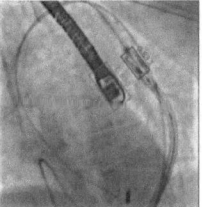

Fig. 5. Center point heatmaps and object detection with confidence values

2.5 Clinical Application: Fusing 3D Echo with Live X-ray Images

The object detector can detect the location as well as in-plane rotation (Fig. 6(a)) and scale of the TOE probe. There are two additional rotations: roll and pitch (Fig. 6b), which are out-of-plane. Roll and pitch angles could not be detected by our object detector. A template library was developed to detect both out-of-plane rotation angles and it is a comprehensive collection of images of the TOE probe in different roll and pitch rotation angles. These images are created from a digitally reconstructed radiography (DRR) model of the TOE probe. As the TOE probe is sitting inside the oesophagus during the procedure, the probe is not free to move in all directions. Our template library only covers the pitch angle from -45° to 45° and the roll angle from -90° to 90°. The angle interval is 2°. Therefore, the number of images in the library is 4050 images ($4050 = (180/2) \times (90/2)$). The normalized cross correlation is used to compute the similarity between the detected probe image patch and an image from the template library. A real-time performance can be achieved by using a GPU-based implementation.

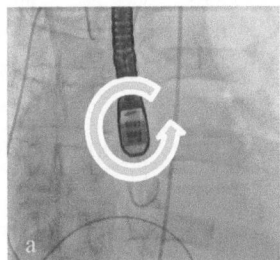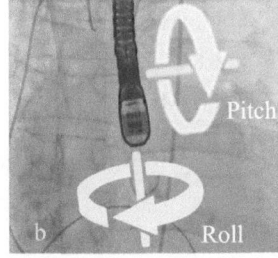

Fig. 6. (a) In-plane rotation. (b) Roll and pitch (out-of-plane). (c) The DRR model

The 3D TOE image volume can be visualized in the 2D X-ray fluoroscopic image by aligning the TOE and X-ray system coordinate systems. The transformation matrix, $T_{TOE\_to\_Xray}$, which transforms from 3D TOE image space to 2D X-ray image space consists of a rigid body transformation matrix T_{rigid} and a projection matrix T_{proj}. It can be computed as $T_{TOE\_to\_Xray} = T_{proj}T_{rigid}$. The projection matrix transforms from 3D X-ray C-arm space to 2D X-ray image space. This can be calculated by using the intrinsic parameters of the X-ray system [21]. T_{rigid} can be decomposed into two matrices ($T_{rigid} = T_{model\_to\_C-arm}T_{TOE\_to\_model}$). Where $T_{TOE\_to\_model}$ transforms from 3D TOE model space to 3D X-ray C-arm space. This matrix is generated by the probe detection algorithm and probe image matching method that positions the 3D TOE model in C-arm space. $T_{TOE\_to\_model}$ relates the position of the 3D TOE images to the position of the 3D TOE model. This is the TOE probe calibration matrix and is calculated pre-procedurally using a specifically designed calibration phantom and the calibration method can be in [22].

3   Results

A total of 10,072 X-ray images (80% train, 10% validation and 10% testing) was used to train and test object detectors. Both validation and testing images are real images, and 4,789 synthetic images were only used in the training dataset. All models using different backbones were implemented in Keras with a Tensorflow (version 2.10) backend and were trained on a GPU farm (NVidia RTX 6000 Ada with 48G memory). The trained models were evaluated on an Intel i7 1.8GHz laptop with a NVidia T550 graphics card to test the inference speed. Tables 1 and Table 2 show the comparison results of our approach (AttWire) with state-of-the-art backbones in single and multiple object detection. AP_{50} and AP_{75} are the average precisions, which are evaluated at IoU $= 0.5$ and IoU $= 0.75$. AP_{50} and AP_{75} in multiple object detection are the mean values of all objects. mAP is the mean value across different IoU thresholds (IoU thresholds from 0.5 to 0.95 with a step size of 0.05).

Table 1. Results for single object detection.

Target Object	Backbone	Parameters	FPS	IoU	AP_{50}	AP_{75}
TOE probe head	VGG16	17.1M	43	0.77 ± 0.21	0.9	0.812
	ResNet-50	36.4M	31	0.83 ± 0.14	0.977	0.794
	AttWire	6.8M	55	0.89 ± 0.05	1.0	0.978
Aortic valve	VGG16	17.1M	52	0.81 ± 0.11	0.972	0.743
	ResNet-50	36.4M	37	0.85 ± 0.08	0.981	0.886
	AttWire	6.8M	59	0.93 ± 0.04	1.0	1.0

Table 2. Results for multiple object detection.

Backbone	Parameters	FPS	IoU (TOE)	IoU (valve)	mAP	AP_{50}	AP_{75}
MobileNet	7.3M	53	0.79 ± 0.09	0.77 ± 0.17	0.546	0.999	0.603
ResNet-50	28.7M	41	0.81 ± 0.07	0.80 ± 0.15	0.618	0.998	0.729
DenseNet121	11.1M	34	0.81 ± 0.07	0.79 ± 0.16	0.584	0.997	0.644
AttWire	3.7M	58	0.88 ± 0.04	0.87 ± 0.11	0.779	1.0	0.922

Overall accuracy of fusing 3D echo with X-ray images was evaluated by using target registration error (TRE). TRE is defined as error distances between corresponding points in both X-ray and echo images. Although real-time synchronized visualization of the live data stream was possible during the clinical procedures, the post-procedure analysis for this paper required that the recorded X-ray and echo data were synchronized manually, resulting in only approximately synchronized sequences. The manual synchronization was done through visual matching using landmarks such as catheters or artificial valves. Total 20 overlay views are created from 10 X-ray image sequences. Corresponding catheters were manually defined in the echo and X-ray views using spline curves. Equally spaced points along the echo curve were automatically defined as measurement points. The corresponding X-ray point was defined as the closest point on the X-ray curve. An example of these error measurements is given in Fig. 7(b). Overall, our method achieves a TRE of 2.5 ± 1.2 mm at a speed of 32 FPS.

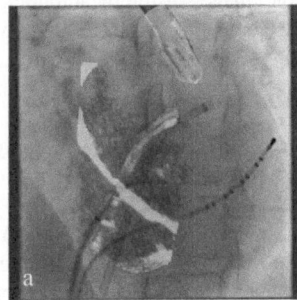 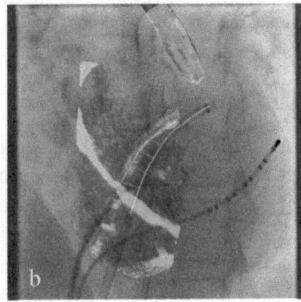

Fig. 7. An example of error measurement. (a) Echo X-ray overlay. (b) Error measurement. Red lines are the shortest distances.

4   Conclusion and Discussions

Clinical applications for minimally invasive heart procedures require highly robust and accurate algorithms for detecting interventional and imaging devices in real-time X-ray fluoroscopic images. In this paper, novel attention mechanisms were designed to guide the CNN model to the areas of wires in X-ray images. The attention-based backbones were implemented in both single and multiple object detection models and they outperform existing state-of-the-art and light-weight backbones by every metric. In addition, our single object detection framework has achieved above 0.97 in AP_{75} and more than 50 FPS. The proposed models for multiple object detection also can perform keypoint detection. With the attention mechanisms we designed, the framework could robustly localize the positions of electrodes on the catheter, and this will enable detecting catheters and devices simultaneously. The detection CNN model facilitates real-time fusion between X-ray fluoroscopy and 3D echo images. It could provide both detection of both TOE probe and other surgical devices.

Acknowledgments. This study was funded by EPSRC, UK (grant number EP/X023826/1).

References

1. Rhode, K., et al.: Clinical applications of image fusion for electrophysiology procedures. In: ISBI 2012, pp. 1435–1438 (2012)
2. Housden, R.J., et al.: Evaluation of a real-time hybrid three-dimensional echo and X-ray imaging system for guidance of cardiac Catheterisation procedures. In: Ayache, N., Delingette, H., Golland, P., Mori, K. (eds.) Medical Image Computing and Computer-Assisted Intervention – MICCAI 2012. MICCAI 2012. Lecture Notes in Computer Science, vol. 7511. Springer, Berlin, Heidelberg (2012). https://doi.org/10.1007/978-3-642-33418-4_4 25–32
3. Ma, Y., et al.: Real-time registration of 3D echo to x-ray fluoroscopy based on cascading classifiers and image registration. Phys. Med. Biol. **66**(5) (2021)
4. Panayiotou, M., et al.: A statistical method for retrospective cardiac and respiratory motion gating of interventional cardiac X-ray images. Med. Phys. **41**(7), 071901–071913 (2014)
5. Mountney, P., et al.: Ultrasound and fluoroscopic images fusion by autonomous ultrasound probe detection. In: Ayache, N., Delingette, H., Golland, P., Mori, K. (eds.) Medical Image Computing and Computer-Assisted Intervention – MICCAI 2012. MICCAI 2012. Lecture Notes in Computer Science, vol. 7511. Springer, Berlin, Heidelberg (2012). https://doi.org/10.1007/978-3-642-33418-4_67
6. Gao, G., et al.: Registration of 3D trans-esophageal echocardiography to X-ray fluoros-copy using image-based probe tracking. Med. Image Anal. **16**(1), 38–49 (2012)
7. Heimann, T., Mountney, P., John, M., Ionasec R.: Learning without labeling: domain adaptation for ultrasound transducer localization. In: Mori, K., Sakuma, I., Sato, Y., Barillot, C., Navab, N. (eds.) Medical Image Computing and Computer-Assisted Intervention – MICCAI 2013. MICCAI 2013. Lecture Notes in Computer Science, vol. 8151. Springer, Berlin, Heidelberg (2013). https://doi.org/10.1007/978-3-642-40760-4_7
8. Hatt, C.R., Speidel M.A., Raval, A.N.: Hough forests for real-time, automatic device localization in fluoroscopic images: application to TAVR. In: Navab, N., Hornegger, J., Wells, W., Frangi, A. (eds.) Medical Image Computing and Computer-Assisted Intervention -- MICCAI 2015. MICCAI 2015. Lecture Notes in Computer Science, vol 9349. Springer, Cham (2015). https://doi.org/10.1007/978-3-319-24553-9_38
9. Miao, S., Wang Z., Liao R.: A CNN regression approach for real-time 2D/3D registration. In: IEEE Trans. Med. Imag. **35**(5) 1352–1363 (2016)
10. Marc, D., et al.: ConTrack: contextual transformer for device tracking in X-Ray, In: MICCAI 2023, LNCS, vol. 14228, pp 679–688 (2023)
11. Arkin, E., et al.: A survey: object detection methods from CNN to transformer. Multimedia Tools Appl. **82**, 21353–21383 (2023)
12. Long, W., Yu, C., Yi, F., Xinyu, L.: A comprehensive survey of oriented object detection in remote sensing images. Expert Syst. Appl. **224**(4), 1–16 (2023)
13. Pérez, P., Gangnet, M., Blake, A.: Poisson image editing. In: ACM SIGGRAPH 2003, pp. 313–318 (2003)
14. Tan, J., Hou, B., Day, T., Simpson, J., Rueckert, D., Kainz, B.: Detecting outliers with poisson image interpolation. In: MICCAI 2021. LNCS, vol. 12905, pp. 581–591 (2021)
15. Frangi, A.F., Niessen, W.J., Vincken, K.L., Viergever, M.A.: A Multiscale vessel enhancement filtering. In: MICCAI 1998. LNCS, vol. 1496, pp 130–137 (1998)
16. Ma, Y., Alhrishy, M., Narayan, S.A., Mountney, P., Rhode, K.S.: A novel real-time computational framework for detecting catheters and rigid guidewires in cardiac catheterization procedures. Med. Phys. **45**(11), 5066–5079 (2018)
17. Shen, Z., Zhang, M., Zhao, H., Yi, S., Li, H.: Efficient attention: attention with linear complexities. In: IEEE Winter Conference on Applications of Computer Vision (WACV), 3530–3538 (2018)

18. Qian, W., et al.: Learning modulated loss for rotated object detection. In: Proceedings of the AAAI Conference on Artificial Intelligence, vol. 35, pp. 2458–2466 (2021)
19. Zhou, X., Wang, D., Krähenbühl, P.: Objects as points. arXiv preprint arXiv:1904.07850 (2019)
20. Lin, T., Goyal, P., Girshick, R., He, K., Dollár, P.: Focal loss for dense object detection. IEEE Trans. Pattern Anal. Mach. Intell. **42**(2), 318–327 (2020)
21. Hawkes, D.J., et al.: The accurate 3-D reconstruction of the geometric configuration of vascular trees from X-ray recordings. Phys. Eng. Med. Imag. **119**, 250–256 (1987)
22. Housden, R.J., et al.: Spatial compounding of trans-esophageal echo volumes using X-ray probe tracking, pp. 1092–1095 (2012)

Author Index

K. Drechsler et al. (Eds.): CLIP 2024, LNCS 15196, pp. 95–96, 2024.
https://doi.org/10.1007/978-3-031-73083-2